CHANGING GEOGRAPHY

SERIES EDITOR: JOHN BALE

'side conflicts

RICHARD YARWOOD

Geographical Association

ACKNOWLEDGEMENTS

The author would like to thank John Bale and the editorial team at the Geographical Association for their comments and advice on drafts of this book.

I am grateful to Professor Brian Ilbery and Carol Blower for information about Hatton Country World in Case Study 1; the Welsh Joint Education Committee for permission to reproduce one of their examination questions in Activity Box 8; Grenville Sheringham for the LEADER exercise in Activity Box 13; Andy and Matt Maginnis of Worcestershire County Council's Countryside Service for the information about Hartlebury Common (Activity Box 19); and National Paintball Games for permission to use Figure 24.

This book is dedicated to my parents.

AUTHOR: Dr Richard Yarwood is a Lecturer in the Department of Geographical Sciences at the University of Plymouth.

© Richard Yarwood, 2002

This book is copyright under the Berne Convention. All rights are reserved. Apart from any fair dealing for the purpose of private study, research, criticism or review, as permitted under the Copyright, Designs and Patents Act 1988, no part of this publication may be reproduced, stored in a retrieval system, or transmitted in any form or by any means, electronic, electrical, chemical, mechanical, optical, photocopying, recording or otherwise, without the prior written permission of the copyright owner. Enquiries should be addressed to the Geographical Association. The author has licensed the Geographical Association to allow, as a benefit of membership, GA members to reproduce material for their own internal school/departmental use, provided that the author holds copyright. The views expressed in this publication are those of the author and do not necessarily represent those of the Geographical Association.

ISBN 1 84377 001 6
First published 2002
Impression number 10 9 8 7 6 5 4 3 2 1
Year 2005 2004 2003

Published by the Geographical Association, 160 Solly Street, Sheffield S1 4BF. The Geographical Association is a registered charity: no 313129.

The Publications Officer of the GA would be happy to hear from other potential authors who have ideas for geography books. You may contact the Officer via the GA at the address above.

Edited by Rose Pipes
Designed by Arkima Ltd, Leeds
Printed and bound at Stanley Press, Dewsbury

CONTENTS

		EDITOR'S PREFACE	4
		INTRODUCTION	5
CHAPTER	1.	WHERE IS THE COUNTRYSIDE?	7
	2.	A PLACE OF WORK	13
	3.	THE GOOD LIFE	23
	4.	BEYOND THE CHOCOLATE BOX	32
	5.	A PLACE TO PLAY	43
		REFERENCES AND FURTHER READING	56

EDITOR'S PREFACE

The books in the *Changing Geography* series seek to alert students in schools and colleges to current developments in university geography. It also aims to close the gap between school and university geography. This is not a knee-jerk response – that school and college geography should be necessarily a watered-down version of higher education approaches – but as a deeper recognition that students in post-16 education should be exposed the ideas currently being taught and researched in universities. Many such ideas are of interest to young people and relevant to their lives (and school examinations).

The series introduces post-16 students to concepts and ideas that tend to be excluded from conventional school texts. Written in language which is readily accessible, illustrated with contemporary case studies, including numerous suggestions for discussion, projects and fieldwork, and lavishly illustrated, the books in this series push post-16 geography in the realm of modern geographical thinking.

Countryside conflicts enables students to investigate the countryside from a range of perspectives – as a place of work and a place to play. It will provide students with a wealth of information to study the economic and social consequences of government policies, and of people's attitudes to, and conceptions of, the countryside. The contents of this book will also be of use to students following courses in Sociology, Leisure and tourism and Planning.

John Bale
February 2002

INTRODUCTION

There has never been so much interest in the countryside. The shelves of newsagents are full of magazines extolling the virtues of rural life, and membership numbers for countryside and conservation organisations are at record levels. The countryside provides a welcome respite from the stress of urban living and has long been popular as a place to visit and to retire to. However, for those whose livelihood depends on the countryside, there are many problems and hardships to face. For example, the foot and mouth epidemic of 2001 led to the destruction of livestock, despair for farmers, the closure of the countryside and a loss of income for thousands. Even before the outbreak, many people living in the countryside had participated in fuel protests, believing that transport costs were destroying country businesses, and many more had protested against plans to ban hunting with hounds, complaining that it would destroy rural life. Away from these high-profile issues, drug-abuse, homelessness, crime, poverty, low wages and isolation are significant, but hidden, problems in many rural areas.

There are so many demands and pressures on rural space that it is impossible to understand them from just one perspective. It is little wonder that rural issues are a key part of geography courses in further and higher education. Whether from the global viewpoint of a European policy maker or the alternative standpoint of a New Age traveller, geography provides the ideas and methodologies to allow a holistic understanding of all walks of rural life.

This book examines the countryside through the eyes of a geographer. Emphasis is placed on 'ways of seeing' the countryside and the approaches that are available to help understand the society, economy and culture of rural areas in the UK. As well as becoming more aware and better informed about rural issues, it is also hoped that the range of topics and approaches covered in the book will help you to appreciate the value and breadth of current geographical thought. To aid your learning, Information Boxes have been used to highlight specific issues in detail, and Activity Boxes offer practical activities designed to help your thinking on specific ideas.

POST OFFICE STORES
ST GILES ON THE HEATH
EDWARD & NADINE TITCOMBE

THE PINT
&
POST
TEA ROOM & BAR

CHAPTER 1

WHERE IS THE COUNTRYSIDE?

We probably all have a fairly good idea about what a rural area is. Spend a few moments studying the photographs in Figure 1 and look at the questions in Activity Box 1 and then decide which ones show the countryside.

Although the questions in Activity Box 1 seem quite straightforward, geographers and planners have spent much effort attempting to define 'rural' and 'urban' areas. Despite this, there is still no universal agreement on what is meant by these terms.

Consider how you approached the activity. What criteria did you use to distinguish between urban and rural places? First, you may have considered the type of *landscape* in the pictures: rural areas are often distinguished as places that are 'wild' or 'natural'. Or you might have considered the kinds of *economic* activities taking place: perhaps the countryside is most often associated with farming, forestry or an apparent lack of economic development.

Activity Box 1: Rural landscapes?

Examine the photographs shown in Figure 1.
1. For each area shown, do you consider it to be 'rural', 'urban' or 'both'?
2. Are some areas more rural than others?
3. List the criteria you have used to make your decisions.

Figure 1: Rural areas, urban areas, or both?

You may also have considered the *built environment*. Is a rural area one that has fewer buildings than an urban or a suburban area? If this is the case, is it possible simply to draw lines around areas of high building density and call them urban? Or should we consider the *interaction* between urban and rural places? After all, many people live in the country but work in the city, and many people living in the city use the countryside for leisure and recreation. Rural areas provide resources and food for urban areas, while urban areas provide administration, services and facilities for rural areas. These interactions are most apparent in the *urban fringe*, but all settlements provide services for their *rural hinterlands* and these vary with the size of the settlement. We might expect to see one major city providing higher-order services (such as major shops, hospitals, seats of local government and so on) for an entire region. Beneath this in the hierarchy may be a few towns providing middle-order services (such as markets, schools, doctors) for a rural district, then, beneath these in the hierarchy are many local villages providing lower-order services (such as smaller shops, post offices, pubs) for their immediate population. Perhaps it is better, therefore, to consider places as interactive regions, rather than breaking them down into urban and rural areas in a rather artificial way.

Or you might also have considered *isolation*. Is an area that is further away from a city or town more rural than one that is closer? If this is the case, how does Britain compare with other countries? Can we really describe any part of Britain as 'rural' compared with the vast tracts of wilderness in, say, Australia or South Africa? What we consider to be rural is therefore relative.

You might also consider the *people* and *communities* of urban and rural areas. Rural areas are often defined as having low population densities, although these definitions vary from country to country. In Britain areas of less than 10,000 people are often classed as rural, though the figure is 250 in Norway! Is there a different *lifestyle* associated with urban and rural areas? It is widely held that rural communities are more closely knit than those in urban areas and it is often thought that life is less stressful, more laid back and problem-free in the countryside. Some speak of country 'ways of life' that are not appreciated or understood by urban people. But what makes a 'country' person? Is it someone who farms, shoots, fishes and hunts? What about people living in the countryside who do none of these things? Are they 'true' country people? Or is it necessary to live in the country for a certain amount of time before one is considered 'truly rural'?

As well as people, *animals* and wildlife might also come into our thinking. We might associate rural areas with farm animals; the presence of cows and sheep in fields are an important part of rural identity. We certainly do not expect to see cows wandering around a British city! Rural areas are also important habitats for wild animals, birds and plants. Indeed, many people feel that rural areas should be *protected*, and some areas have been given a special designation to enable planners to do this. Are these rural areas more valuable than others? Similarly, other (and sometimes the same) parts of the countryside have been designated for outdoor *leisure and recreation*. National Parks and National Trails are good examples. What do these designations reflect? How do they help us to distinguish between different rural areas?

Are rural areas really that *different* from the city? How would you classify a park in a city? Does the fact that it is open, designated for recreation and often has an abundance of wildlife make it a rural place? Likewise are there 'villages within cities' with closely-knit communities? And, conversely, are commuter villages in the countryside very 'rural' if most of their residents work in the city?

As you will be starting to appreciate, what is at first a simple question provokes a great deal of thought. There are many ways that we can distinguish between urban and rural places, and between different rural areas. But does it really matter? After all, we all have a good idea about what is or is not the countryside! Why can't we just 'get on with it' and study the countryside?

For gegraphers, it is important to consider *'rurality'*, or what is meant and understood by the term 'rural', as this has determined how rural areas have been studied. It is also important to understand what ordinary people understand and value about the countryside. Although there are some widely held views about the countryside, there are also many different, opposing ideas and these have led to conflict in rural places. Thus, it is important to consider these 'lay' views as well as more academic definitions.

Mapping rural places

One way to overcome some of the dilemmas associated with defining urban and rural areas, is to take a 'scientific' approach to the problem. Geographers and planners have gathered relevant data, fed them into software packages and used them to define and map rural areas. This approach was seen to be objective as it defined urban and rural areas according to scientific models. Some examples of these approaches are given in Information Box 1, together with the criteria used in the model.

CHAPTER 1: WHERE IS THE COUNTRYSIDE?

Information Box 1: Mapping rural areas

Maps of rurality and the criteria used to create them, including:

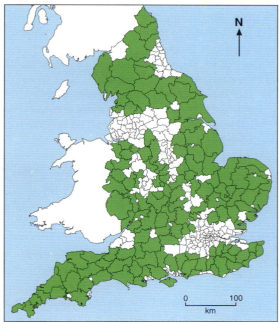

Figure 2: The Countryside Agency's landscape areas. Different parts of rural England have their own 'character' which has come about from a combination of natural and human processes. The information is based on landscape, wildlife and natural features to show unique rural environments.

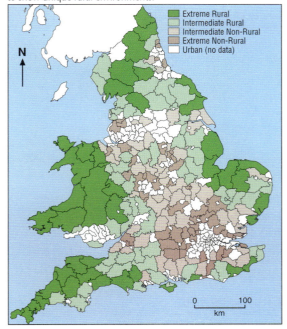

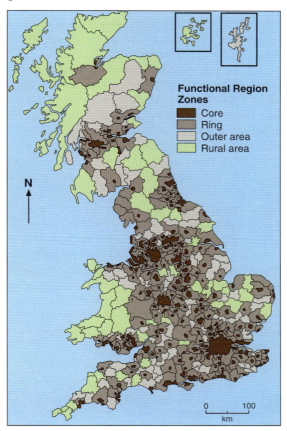

Figure 4: Functional areas: CURS classifications. The maps of Functional Areas was produced by a team of geographers. They were interested in examining the relationship between the town and country. Like the Rurality Index, social and economic data were used to map different kinds of rural area. However, greater emphasis was placed on the economic interaction between rural areas, in particular commuting links. Thus, the functional regions demonstrate the relationship between cities and their rural hinterlands. Source: Champion *et al.*, 1984.

Figure 3: Socio-economic data: 1991 Rurality Index.
The Rurality Index was initally devised by Professor Paul Cloke. In contrast to Figure 2, the Rurality Index uses information based on human social and economic activity, rather than landscape features. Data from the Census are used to distinguish 'how rural' a district is. These include measures of population density and structure, economic activity (including farming and the number of commuters), migration and distance from major urban centres. Source: Harrington and O'Donohugue, 1998.

Information Box 1 indicates that despite efforts to be objective, there is a degree of subjectivity in mapping urban and rural places. This is because geographers and planners have to choose what information they will use and how they will use it. Activity Box 2 will help you to appreciate why these decisions result in different maps.

Despite this 'objective' approach, no two maps are identical. This is because geographers choose different kinds of data as good indicators of a rural area. One may choose to emphasise distance, another the economy, and so on. Although the calculations are carried out impartially, the choice of data is subjective.

The maps also tend to *describe* rural areas. They tell us very little about the human experience of the countryside and *what is it like* to live in or visit. Furthermore, the maps do little to *explain* the differences between rural places and the processes affecting them. Nevertheless, these kinds of 'objective' maps do provide us with a good starting point. They recognise that there is a geography of the countryside and that there are different types of rural space. For planners and other decision makers, these maps and models allow rural areas to be treated with a degree of impartiality. However, some geographers have sought other ways of studying the countryside.

Is there really a difference?

Some geographers argue that to *explain* rural change, we should focus on understanding the social, economic and political *processes* that impact on rural areas. If this is the case, we should spend less time trying to define rurality and more time studying the factors that affect rural society and the rural economy. Perhaps we should focus on the causes, rather than the symptoms, of rural change.

In this sense, some geographers advocate 'doing away' with rurality: in other words studying urban and rural areas in the same manner. This approach can be justified in a number of ways:

- Urban and rural areas have similar *economies*. The service (tertiary) sector is now the biggest employer in many urban and rural localities as 'traditional' industries, manufacturing and agriculture respectively, have declined.

- *Socially*, urban and rural areas share similar problems. Crime, drug abuse, homelessness and social exclusion are not confined to cities and towns. Commuting, teleworking and the mass media have also blurred the distinction between urban-dwellers and rural-dwellers.

Activity Box 2: Distinguishing between urban and rural places

Using data from the most recent Census examine two neighbouring counties, one metropolitan and one non-metropolitan, that you are familiar with. You may wish to use data that you think reflects:

- the economy or employment structure;
- commuting patterns;
- population;
- housing conditions;
- quality of life.

Using these criteria, produce maps that compare and contrast social and economic differences between and within the two counties. It is up to you to decide what scale you map this information and precisely what data you use.

Compare your maps with others in your study group and discuss the following:

- Using your maps can you distinguish between urban and rural areas easily?
- Can you distinguish different types of urban and rural areas with the two counties?
- How helpful are census data for this task? Would you like any more information? Do you feel that anything is 'missing' from your definitions (such as a sense of what the landscape of the areas is 'like')?
- Do your maps differ from others in your class? How significant are these differences? Are they caused by your choice and use of data?
- Speculate on the reasons for the spatial distributions you have observed.

- *Politically*, urban and rural areas are sometimes grouped together. The current Labour government, for example, is emphasising the *interdependence* of urban and rural areas and, consequently, many policies are aimed at strengthening links between city and country.

Given that urban and rural areas have become increasingly blurred, why should we make a distinction between them?

Cultural constructions of rurality

Most of us *feel* that there is a difference between urban and rural areas, we place certain values on the countryside and sometimes behave differently in the town than the country. For example, we may say 'hello' to people we pass on a country walk (Gilg, 1996), but would not do this on a walk in the city centre: we would probably get some very strange looks if we did.

Society, organisations, groups and individual people all have different *ideas* about the country. It is important to understand these because they reflect and affect the way in which rural areas are used. Geographers have tried to understand how places are given *cultural* identity. Culture refers to a 'shared understanding' between different people or groups.

How can we start to study culture? Imagine two crowds at a football match. Both sets of supporters probably come from similar social groups and are of similar age, gender and class. A statistical test on these people would reveal 'no significant difference' between these two groups. But try telling that to the two sets of fans! They identify very closely with 'their' team and no doubt feel very different at the end of the match, depending on who has won or lost. If you were the only person wearing blue in a sea of red, you would feel very 'out of place'!

Clearly, statistics (*quantitative* sources) are of limited value in the study of culture. Instead, geographers have started to use *qualitative* sources to study people's understanding of and behaviour in rural areas. An exciting development in geography has been the study and analysis of media such as films, books, paintings, art, poetry, advertising, television, radio, cartoons and to reveal commonly held ideas about rurality.

One long-standing idea is that the countryside is a far more pleasant place to live than the city. The pace of life is seen to be slower, friendlier and less stressful than in the city (Information Box 2). It is a widely held view that there are fewer 'problems' in rural areas and the environment is viewed as cleaner and more 'traditional'. This idealised view is referred to as the *rural myth* or *rural idyll*.

Information Box 2: Contrasting ideas about town and country (After: Short, 1989).

Rural idyll	Urban nightmare
■ Nostalgic/part of national identity	■ Lacking identity
■ Traditional	■ Modern
■ Problem-free	■ Crime, poverty, homelessness
■ Closely knit/friendly	■ Anonymous/lonely
■ Better environment	■ Urban decay
■ Place of play	■ Place of fear
■ Simpler/more natural	■ Polluted, congested, dirty

Anti-idyll	Urban dream
■ Backward	■ Progressive
■ Unsophisticated	■ International/cosmopolitan
■ Unfriendly/hostile	■ Diverse, freedom to express yourself
■ Environmentally damaged	■ Architectural achievement
■ Dull, boring	■ Exciting, recreational
■ Poorly provided with services	■ Shopping, administrative centres
■ Sleepy	■ 24-hour city

CHAPTER 1: WHERE IS THE COUNTRYSIDE?

Evidence of the rural idyll can be seen in many forms of media. For example, advertisements often use images of the countryside to emphasis the natural, wholesome qualities of various food products. A browse around any newsagent will reveal a whole series of magazines aimed at 'traditional' rural pursuits such as hunting, shooting and fishing. Others are written for recreationalists, showing the to be a place of gentle leisure or challenge. Television series, such as *Heartbeat*, celebrate nostalgic, closely-knit communities of yesterday. Estate agents, too, exaggerate the virtues of rural life in their attempts to sell properties in the countryside.

There is, in may people's minds, a strong link between the countryside and a past, or traditional, way of life for which they feel great nostalgia. The works of the Romantic poets, such as Wordsworth (who lived in and wrote about the Lake District – see page 44), have elevated rural areas to special, even spiritual status, and paintings of the countryside as it was over 200 years ago by artists such as J. Constable or J.M.W. Turner provoke feelings of longing for the apparent simplicity and tranquillity of rural life.

The rural idyll is a dominant idea and, as you will see in later chapters, has strongly influenced the way in which people use and view the countryside. For example, it has encouraged many people to migrate from town to country in search of what they perceive to be a better lifestyle.

However, not everybody shares in the rural myth, and there are many examples of books, plays, films and so on, which present the countryside in negative terms. Compare, for example, the different images of the Irish countryside portrayed in the television series *Father Ted* and *Ballykissangel*. Consider television series such as *The League of Gentlemen* or *Viz* magazine's 'Farmer Palmer', in which rural people are portrayed as backward and hostile. Horror films, such as *The Blair Witch Project*, show the frightening side of rural isolation. Although Laurie Lees' (1959) novel *Cider With Rosie* is often held up as portraying a nostalgic way of life, it contains reports of crime, sexual deviance, poverty and exclusion.

People's views of the countryside vary widely, and tend to reflect their particular relationship with it. A farmer, for example, may see the countryside as 'a factory floor'; the conservationist as a threatened environment; a young person as a dull place with too little excitement; the New Age traveller as a space that allows him/her to drop out of society; the soldier as a training ground; and so on. Clearly, people use and perceive the countryside in very different ways, and as more and more people live in or use the countryside, these differences may lead to conflict. Given this variety in the use of rural space, perhaps it is better to talk about the *geographies*, rather than the *geography* of the countryside.

Many geographers now talk about 'the rural'. Although an English teacher may object to an adjective being treated as a noun, this usage highlights the importance given to the *idea* of the countryside and recognises 'the rural' as a subject in its own right; and one that is constructed, imagined and contested by diverse groups of people.

Geographers have attempted to define rural areas using a range of perspectives. It is important to consider what we mean by a 'rural area', but we should not waste time trying to come up with an all-encompassing definition. Although it is helpful to map different rural areas using different criteria, such maps should be seen only as starting points. Other 'texts' help to understand how the countryside can be defined by different people.

To understand rural areas fully it is important to research the social and economic processes that affect them, and to consider how they are valued and contested by different groups of people. In Chapters 2-5 you are encouraged to consider how these ideas can be used to inform your understanding of rural places as well as the changes taking place in them.

Activity Box 3: A qualitative study of rurality

Choose a novel, film, poem, painting, song, advertising campaign (e.g. holiday brochures, college prospectuses, television or radio advertising), or some other medium that you are familiar with which depicts the countryside or the city, or both.

Discuss the ways in which the city or countryside are represented in the work you chose.

Give evidence from your chosen work to support your answers to the following questions:

- Are attitudes to the country or city positive or negative in the work?
- What devices does the author, painter, producer, scriptwriter, etc., use to convey this attitude?
- Does the work reflect prevailing ideas of the country and city?
- How do you think the work might influence people's views of or behaviour in the countryside or city?

CHAPTER 2

A PLACE OF WORK

Chapter 1 considers the different ways in which geographers have tried to conceptualise and explain the countryside. Some have argued that we should not seek to differentiate between urban and rural areas but, instead, consider how structural processes (especially the economy), span and affect both urban and rural areas. It is important to understand changes in the rural economy if we are to understand changes in rural life and living. This chapter outlines the major changes in rural employment and the implications this has for rural society. It provides a broad overview of change in the context of the UK but, needless to say, not every place has changed in the same way and there are local variations in these trends.

Work has an impact on the social relations within an area. In the past both urban and rural communities were often based around a single industry. Communities were drawn together by the shared activities and the hardship of work and, as a result, local traditions developed. When these single industries closed or changed, there were social and cultural, as well as economic impacts on the communities. Historically, agriculture dominated local employment and tradition in many rural areas. Farming practices and working relationships helped to shape rural communities and create local ways of life. Today, as Figure 5 shows, the picture is very different.

Industry also connects local places to national and global economies. No longer do rural areas simply provide food and raw materials for their urban neighbours. Today, decisions about which crops to grow or which farming practices to follow are influenced more by government or European Union (EU) policy than local farming conditions or local market forces. Similarly, the importing of raw materials and the exporting of finished products from rural manufacturing plants connects villages and towns to people and economies in places far beyond the local area. In short, the rural economy has become *globalised*. To understand local employment change and, in turn, social relations, it is important to consider broad economic changes and their impacts on particular places (Figure 6).

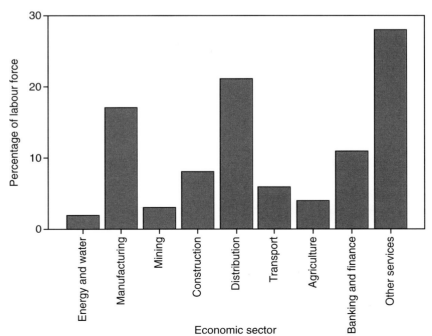

Figure 5: Employment by economic sector in rural England, 1991.
Source: Census 1991.

CHAPTER 2: A PLACE OF WORK

As ways of work in rural areas have changed (Figure 5), and as Information Box 3 shows so have ways of life. In general terms, the rural economy has moved from one based on *production*, to one based on *consumption*. Production refers to the making of goods or growing of food, while consumption refers to the buying and selling of goods or services.

Agriculture

Agriculture plays an important role in the environment and society of the countryside. Farming practices have shaped, and continue to shape, rural landscapes. 'Arable land' and 'managed grasslands' still account for nearly two-thirds of land use in rural England. Although only 600,000 people are directly employed in agriculture in the UK (about 2% of the total work force of the UK), there are 3.3 million jobs associated with the UK food chain (DETR, 2000). However, the farming industry has undergone dramatic change in the post-war years. Ilbery and Bowler (1998) have suggested that this can be characterised by two main phases: *a productivist phase* and *a post-productivist phase*.

Productivist phase

Productivist farming (practiced from around 1945 to the mid-1980s) aimed to make Britain self-sufficient in food production after the shortages experienced during the Second World War. Farmers were offered subsidies to specialise in certain types of food production, use more machinery, buy more products to feed livestock and to use more artificial fertilizers. To achieve economies of scale, farms became more *specialised* (Table 1) and *intensive* as exemplified by the increased use of inorganic pesticides, greater mechanisation and technology to produce more food from the same, or less, land.

Although they improved food output, these productivist methods caused environmental damage and led to a 'farm crisis', which was due to overproduction of foodstuffs, the immense costs of agrarian support and falling farm incomes. In the mid-1980s these pressures led to the adoption of new farming methods, known as 'post-productivism' (Yarwood and Evans, 1999).

Post-productivist phase

Post-productive farming (mid-1980s onwards) has been characterised by *diversification* and *extensification*. Many farmers have diversified their activities to gain extra income. Agricultural diversification has led to farms growing new products (perhaps organically) so that their farms are less specialised or reliant on single-product markets. In other cases, farmers have diversified their activities into non-agricultural ways, such as offering bed and breakfast accommodation, creating farm parks and providing facilities for adventure sports and other visitor attractions (see Case Study 1).

Figure 6: Economic and social relations: the impact of the global on the local.

Table 1: Specialisation of farming, 1967 and 1988. Source: Britton, 1990.

	1967	1988
Average farm size	34ha	64ha
Number of farms growing cereals	172,000	87,000
Average area in which cereals grown	22ha	45ha
Number of farms with dairy herds	132,000	48,000
Average size of dairy herd	24 cows	61 cows

Information Box 3: The impact of the changing rural economy on local social relations

Decade	Main features of rural economy	Impact on production	Impact on consumption	Employment change	Social impacts
1950s and earlier	Agriculture is dominant in many places. Fishing, mining and quarrying have local importance.	Economy is dominated by food production.	Limited – countryside is mainly a place of production, especially after the Second World War.	Agricultural employment dominates.	In many places, rural life is centred on farming practices and traditions.
1960s	Agriculture still dominates and starts to become more intensive as farmers follow the 'technological treadmill'. Places on the urban fringe experience in-migration and increased commuting.	Food production becomes more intensive. Some villages are turning into 'dormitories' for those who work in urban areas.	Some houses on the urban fringe are bought by commuters.	Agricultural employment is important, but mechanisation leads to job losses and depopulation. The number of commuters living on the urban/rural fringe increases.	As employment opportunities decline, some younger people migrate to urban areas for better opportunities. The character of villages on the urban fringe starts to change as not everyone is employed in the same industry.
1970s	The position of agriculture is threatened as the dominant mode of production as manufacturing branch – plants move to some rural places. More and more people want to live in the countryside. This trend continues into the 1990s.	The increased industrialisation of agriculture. Manufacturing industries increase in the countryside.	The growth of agribusinesses links agriculture more closely to urban markets. Manufacturing branch – plants link some rural areas to wider global economies. Local houses are bought (consumed) by new, middle-class populations.	Losses in farming jobs. Increases in manufacturing jobs. Growth in number of professionals who still work in urban places.	The loss of younger people continues. New manufacturing plants create some job opportunities. Women start to enter the workforce in larger numbers. Counter-urbanisation affects all parts of rural Britain. Many move to seek a 'good life' (Chapter 3).
1980s	Manufacturing and service industries dominate the rural economy. Farmers are encouraged to diversify away from food production.	Rural areas are the only places in the UK which experience an increase in manufacturing production. Agriculture enters a 'post-productive' phase.	Teleworking and electronic communications allow some services to be centred in and sold from rural places. Many rural places are commodified to encourage leisure and tourism, and the sale of houses and other 'rural goods'	Manufacturing employment and service employment increases. Growth in service employment linked to leisure.	Local people find it harder to live in their home areas as they are unable to compete in local housing markets. The number and influence of new people in all parts of the countryside increases.
1990s/ 2000	Service sector dominates rural employment.	Since the 1950s most rural areas have gone from being places of production to places of consumption.	Emphasis is very much on the commodification of rural areas to encourage consumption.	Service employment dominates most rural areas. Agricultural employment is in a minority. Manufacturing shows some job losses.	Rural areas are diverse places. Many people live in the countryside but work elsewhere. There are higher numbers of retired people living in rural areas.

Case study 1: Hatton Country World

Hatton Country World is a 283ha (700 acre) estate in Warwickshire that has undergone many of the transformations associated with the 'post-productivist' shift in agriculture. Originally, it was a traditional mixed farm with sheep, beef cattle and arable crops. In the early 1980s it faced financial problems and so, with the aid of various government grants, Hatton Farm started a programme of diversification that was to change its look and business completely. Professor Brian Ilbery of Coventry University has charted these changes in detail.

1982-85
Redundant farm buildings were restored and converted into 18 workshops that were rented to local craft industries. In turn, Hatton promoted itself as a craft centre in order to attract visitors to the industries and to the Farm's 'pick your own' enterprise. A nature trail was also established to encourage visitors. All the sheep were sold and the Farm began to specialise in arable production, with some beef cattle.

1986-88
The craft workshop expanded to 30 units that were accommodated on a specialist site. A café and shop opened in 1987.

1988-89
Thirty-five per cent of (mainly poor-quality) arable land was 'set aside'. Some of this was used as a caravan and camping site. The remaining 65% of arable land was let to two local farmers. Thus, there was no land in arable production on and for the Hatton Estate. A decision was also made to give up beef farming, releasing a 929 sq m (10,000 sq ft) indoor beef complex to be converted to a garden centre and pet shop. The number of craft units was increased to 35. Elsewhere on the Estate redundant buildings were converted to a 15-unit business centre.

1990-93
A Farm Park was opened in 1991. This contained rare breeds of farm animals, a guinea-pig village and nature trail (Figure 7). The garden centre was converted to retail units.

Figure 7: Hatton Country World: farm park.

CHAPTER 2: A PLACE OF WORK

Case study 1: Hatton Country World *(continued)*

1994 onwards
A new restaurant (the Greedy Pig!) was opened together with a function room and children's soft play area. Improvements and new attractions added to the Farm Park. Two 'factory' outlets, selling glass and kitchenware and clothing, were opened on the site. Other recent attractions include an 'Amaizing maze' and various games and activities (Figure 8).

Hatton exemplifies the model of rural economic change described in Information Box 3. Most of its income and employment is based on retailing, leisure and tourism, rather than growing food. It has shifted successfully from production to consumption. A key feature of this change has been the commodification of the farm: agricultural and rural heritage have been used to sell 'Hatton Country World' so that visitors feel that they are buying into rural produce and experience. Most farms have not diversified to the extent of Hatton Country World, but many have become 'commodified' to appeal to a wide range of consumers.

Figure 8: Hatton Country World information board listing additional activities for children of all ages.

Activity Box 4: Farm diversification

Visit a local tourism information office or use the internet to find out about the range and number of farm attractions in a rural area that you are studying. As well as farm parks, look for bed and breakfast accommodation, farm visits, use of farms for sporting or leisure activities, café and restaurants, farm shops, educational activities, walking activities and any other forms of diversification. Using information leaflets and web pages note the following:

- What are the main kind of activities that farmers have diversified into?
- Who are these activities aimed at? Are any particular ideas about rurality (e.g. tradition or heritage) used to promote these activities to particular groups of people?
- Is there a geography of diversification? Do some areas show more signs of diversification than others? Why?
- Do you feel that further diversification is possible in your region? If so, what kind of activities could farmers undertake?

Visit one of the farm attractions you have found out about. How does it compare to Hatton Country World? To what extent has it diversified? How much does it rely on 'traditional' farm activities in relation to provide income and employment? If possible, talk to the farmer about how successful he or she feels the venture has been, how it was funded and what obstacles, if any, hinder progress.

Some farmers have also established other part-time enterprises not directly connected to their farm, such as mobile catering facilities for shows or events in the countryside. In a recent survey of farmers in England and Wales, 56% said that they were operating a non-farming enterprise (Countryside Agency, 2001).

Grants from government and the EU have been made available to help farmers work in extensive, environmentally friendly ways. These have included schemes to 'set-aside' land or take it out of agricultural production. In England, the 'Countryside Stewardship' Scheme encourages landowners to conserve wildlife and landscape and to improve public access to their land. There has also been a growth in organic farming to meet new consumer demands (see e.g. Figure 9). The number of organic producers in the UK rose from around 500 in 1995 to around 2000 in 2000 (Countryside Agency, 2000).

In reality, the shift from productivism to post-productivism does not always happen in the ways described above. Some farms, especially in eastern England, continue to produce food in an intensive fashion. The introduction of genetically modified (GM) crops demonstrates that intensive, scientific production of food is still important. Furthermore, although some farmers may have diversified and adopted environmentally friendly practices on parts of their farms, the greater part of their income may still come from the intensive production of food.

The current farming crisis

Farming in the UK is currently in crisis. In 2000 the total income from farming in the UK fell to its lowest level for over 25 years (to £1.88 billion, compared to £5.5 billion in 1976 (Countryside Agency, 2001)). This has been attributed to:

- the falling prices of milk and cereals;
- the strength of the pound sterling;
- the cost of fuel and fertilisers as well as rising interest rates;
- poor weather (especially the wet autumn and winter in 2000);
- a fall in consumer confidence caused by an increase in, for example, salmonella, bovine spongiform encephalitis (BSE), and foot and mouth disease (see Case Study 2 and Activity Box 5);
- cheaper food imports from overseas.

There have been widespread social consequences. In England, there was a 6% loss of agricultural employment. This was felt most strongly by hired labour, which declined by 12-13%. More work is now being undertaken by farming contractors and 90% of farms in England and Wales employ short-term contractors for at least one job (Countryside Agency, 2001).

Government and EU policy continues to focus on farm diversification, but there are concerns that there are limits to this. Is there enough tourist demand to sustain a visitor attraction on every farm?

Figure 9: Extensive free-range pig farming intended for organic supply.

Diversification also suits larger farms because they are better placed to invest capital and resources into what might be a risky new venture. Diversification is not the only way forward, however. A recent initiative, which has been very successful in the UK, is Farmers' Markets. These are held in towns and cities around the country and enable farmers to sell local produce directly to the public. The first of these markets was held in 1997 and there are now an estimated 300 in the UK. HRH Prince Charles has also called on supermarkets to sell more local produce. Both of these ideas aim to provide local niche markets for farmers, with the emphasis on food quality and local produce, rather than quantity from further away.

The farming recession is causing stress to many farmers. A recent survey found that over 20% were worried about their future but did not know what to do. According to the Samaritans, farmers are one of the occupational groups most at risk from suicide (only vets – another largely rural occupation – and the medical profession are at higher risk). Between 1991 and 1996 there were 190 farming-related suicides in the UK, or one every 11 days. Tragically, this figure may well have increased because of the foot and mouth crisis. The Rural Stress Information Network (www.ruralnet.org.uk/~rusin/) was established to offer help to farmers in stressful and suicidal situations.

Case study 2: The foot and mouth outbreak, 2001

The closure of Dartmoor National Park to the public during the foot and mouth outbreak.

The 2001 foot and mouth disease epidemic has highlighted the importance and fragility of the rural economy. The Department of Food, Rural Affairs and Agriculture state that the outbreak started in February 2001 on a farm at Heddon-on-the-Wall, Northumbria, from where the virus spread to seven other farms in the region. Infected sheep from one of these farms were sent to Hexham Market on 13 February and, from there, were dispersed all over the country. Between 14 and 24 February infected sheep were sent to Longton, Carlisle and Dearham in Cumbria; Highampton in Devon; Lockerbie in Dumfries and Galloway; Nantwich in Cheshire; and, indirectly, to Hatherleigh, Hereford, Northampton and Ross-on-Wye.

The disease spread rapidly in these areas and, by August 2001, 1928 cases had been identified. The areas most severely affected were Cumbria (832 cases), Dumfries and Galloway (176 cases), Devon (173), North Yorkshire (131), County Durham (92), Gloucester (76) and Powys (66) (DEFRA, 2001). In order to maintain meat export markets, the government embarked on a programme of slaughter to contain and eradicate the disease. By August 2001, nearly 3,370,000 animals had been slaughtered.

The movement of people and animals was also highly restricted to prevent the spread of the disease, and the local and general elections which were scheduled for May 2001 were postponed until June. Footpaths were closed in many parts of the countryside and tourism dwindled, as people stayed away.

The causes, impact and control of the disease are still being discussed and analysed, but what is clear is that the outbreak had a significant impact on life and work in the countryside. Over £2 billion has already been spent on the disposal of livestock and clean-up of farms. Table 2 shows the results of an early analysis of the situation in Devon, carried out by the County Council.

Table 2: The effects of foot and mouth on the rural economy in Devon. Source: Turner, 2001.

	Amount of income lost	Number of jobs lost
Agriculture	£114,372,000	1555
Tourism	£107,542,000	3332
Other activities	£94,523,000	2918
Total	£316,437,000	7805

Activity Box 5: Foot and mouth

Look at the statistical information about foot and mouth disease on the Department of Food, Rural Affairs and Environment's (DEFRA) website (www.defra.gov.uk). The site contains a county-level map of foot and mouth disease in Britain and a list of all registered cases in the UK with the date of infection and the location of the farm.

You can use this data for a number of activities, for example:

- to map the spread of the disease across Britain
- to map the distances that animals are transported in intensive farming systems
- to examine the impact of the disease in a particular county.

Many local authorities also produced reports (often available online) that charted the impact of the disease on their county's economy.

Use data from the DEFRA website, reports from local authority websites and survey data to produce a project that examines the impact of foot and mouth disease on a farmer, an owner of abatoir, an owner of a food processing plant, a local butcher, a large supermarket chain and a consumer.

Manufacturing

Manufacturing has, perhaps surprisingly, been a growth area in the rural economy. Gudgin (1995) estimated that between 1960 and 1991 there was a 45% increase in manufacturing jobs in rural areas of Britain, while urban areas lost over 60% of theirs. There are five explanations for the growth of rural manufacturing (North, 1998), as set out below.

1. Government intervention

The Rural Development Commission (RDC) was established in 1910 to encourage economic development in rural areas. During the 1960s and 1970s there was a policy of building 'advanced' factory units, the idea being that these would provide the premises needed for growing rural businesses. The units are generally located on the edge of small towns and large villages although, more recently, there has been a move to convert redundant farm buildings into business premises. In 1996 the RDC was replaced by Regional Development Agencies (RDAs). Regional and national development agencies continue to provide business advice and support for rural enterprises.

2. Room to expand

It has been argued that rural sites offer more space for expansion than urban sites, which may be hemmed in by existing roads and buildings. Rural sites are therefore attractive for growing companies.

3. 'Green' labour force

Some geographers have argued that rural workforces are more attractive to capitalists than urban ones. This is because rural workers are relatively new ('green') to manufacturing employment and, in consequence, they are not well organised to demand improvements in pay and conditions. Thus, the argument goes, capitalists who locate in rural areas are able to exploit this workforce to their advantage (profitability). Some companies have established 'branch plants' in rural locations to manufacture some or all of a product, but their headquarters are often situated in urban areas.

4. Cheaper costs

Related to the above, it may be cheaper to produce goods in rural places. Although transport costs may increase with distance from markets, land rents and labour costs may be cheaper in the countryside.

5. Better environment

Some entrepreneurs choose to locate in rural places in order to work and live in a pleasant environment. This 'countryside asset' is promoted by local authorities and development agencies to attract businesses and investment. There has also been a growth in craft and heritage industries in rural areas: a rural location can add value to craft products by making them appear more 'traditional' or sustainable.

The growth of rural manufacturing has introduced new employment opportunities for rural people but work tends to be low-grade and mundane with few training opportunities or prospects of advancement. Senior positions are often filled from outside the local area and, in some cases, employees are bussed in from nearby settlements – as Case Study 3 illustrates.

Case study 3: The growth of Leominster Industrial Estate

Traditionally, Herefordshire relied strongly on agricultural employment, but in the 1960s and 1970s this started to decline. Local authorities and the Rural Development Commission looked to develop industrial estates in local market towns to provide alternative sources of employment. Leominster, a market town of nearly 10,000 people (1991 Census), was chosen as a site for one of these estates. Between 1971 and 1991 the number of companies nearly doubled, from 25 to 48. Of these, 24 had relocated from elsewhere, nine were branch plants and only nine were new companies. The majority of companies (27) were involved in manufacturing. A survey published in 1996 estimated that 944 people were employed on the estate (Table 3).

Vacancies for manual jobs were generally filled by local people (although nearly half of the companies interviewed had problems with recruitment), while the more skilled (management) jobs were filled by people from outside the region who moved to their posts. The companies offered little training; chose either to employ workers that were already skilled or offered limited on-the-job instruction to manual workers. Thus, while the estate could be seen as successful in creating jobs, it did little to re-train the workforce or to restructure the local labour market in any significant way, and opportunities for local skilled workers remained limited.

Source: Yarwood, 1996.

Table 3: Employment on Leominster Industrial Estate by type of company and occupation.

Type of company	Management		Clerical		Manual		Total
	Male	Female	Male	Female	Male	Female	
Manufacturing	60	14	17	57	295	196	639
Other	41	20	0	15	224	5	305
Total	101	34	17	72	519	201	944

The service industry sector accounts for most (60%) of the jobs in rural England (Figure 5). The service sector includes a wide range of businesses including banking, finance and legal services, education, health and social care and tourism-related businesses as well as shops. Despite many closures (see Chapter 4), these services continue to provide employment for many in rural places. As Chapter 5 shows (pages 43-51), there has been a growth in the range and diversity of leisure activities in the countryside and tourism makes a significant contribution to the rural economy. It has been estimated that visitors to the countryside support 380,000 jobs and 25,000 rural businesses, spending £8 billion a year in England alone (DETR, 2000). Many of these small businesses are important to local entrepreneurs. However, the jobs associated with tourism are often seasonal, low-paid or part-time, and some can be filled from people outside the area (e.g. students doing summer jobs).

Many people live in the countryside but continue to work in urban areas. As mentioned earlier, such people tend to be those with management positions, often in the service sector. Commuting distances have grown as access to private transport and motorway links have improved. Teleworking, e-commerce and the internet are already having an impact on rural areas: they allow more people to work from home using a fax machine, e-mail, the internet and a telephone. Some companies have used this technology to employ people in their homes and to create opportunities for rural workforces. There have also been efforts to connect rural places by providing telecentres and telecottages in the hope that electronic information may overcome the problems of distance and isolation that affect the provision of some services. Advocates of this trend suggest that it improves productivity and reduces traffic congestion. However, it blurs home and work space, encouraging people to work ever-longer hours. Many people miss the companionship and interaction of working alongside others; however, it remains to be seen just how significant the impacts of teleworking are in rural places.

CHAPTER 2: A PLACE OF WORK

Activity Box 6: Rural industrial estates

Using a copy of the *Yellow Pages* for your chosen region (an electronic version is available on www.yell.com) select examples of manufacturing industries (e.g. tool makers, electronic factories, craft industries) and map their location. Further information on local industrial activity might also be available from Local Chambers of Trade, local authorities or Regional Development Agencies.

Use these data to consider:

- What kind of manufacturing activity is occurring in your region? Is it traditional, high-technology or linked to other areas of the rural economy, such as food processing?
- Where are these companies actually located? Are they in major urban areas, small towns, villages or open countryside? Do you think the location might cause problems for employers or employees of these companies?

You can also use information from *Yellow Pages* to map other economic sectors, including tourism or the location of rural shops and services. Compare the distributions of these sectors to that of manufacturing.

Undertake a questionnaire or interview survey of a local industrial estate. The information in Case Study 3 was based on 'cold calling' at businesses in Leominster, which proved to be very successful. Many business people were pleased to have a few minutes off work to answer questions! Local entrepreneurs were especially keen to talk about their businesses.

As well as more general questions about employment (e.g. How many people does the business employ? How many are full-time or part-time? How many are men and how many women? What are the strengths and weaknesses of the local workforce?), you might consider asking:

- Where the company originated? Is it local or has it moved into the area?
- If it moved into the area, why did it choose this particular location?
- What are the advantages and disadvantages of being based in a rural area?
- How have the businesses fortunes fluctuated over time?

Data from this activity can be combined with that from Activity Box 7 to build up a profile of the economy and employment in a particular part of the countryside.

Activity Box 7: Employment survey

Choose a rural area and examine its employment profile. This can be done at a variety of levels using Census data supplemented by job advertisements in the local press and job centres (although the latter tend to be located in urban areas).

Using these data, attempt to discover answers to the following questions:

1. What is the dominant employment?
2. What kind of work is available for local people?
3. How much does it pay? Is it seasonal, part-time, etc?
4. Will travel be an issue for applicants for specific jobs? Does a lack of transport restrict opportunity?
5. How do you think employment, unemployment or underemployment impacts on the area?

The countryside is a working environment for many but, over recent years, the nature of rural work has changed. Consequently communities are no longer centred around agricultural work as many different industries are competing for labour and space. As Chapters 3 and 4 show, these interests can conflict with each other as well as with other demands for rural space (see Chapter 5). As this chapter has shown changing employment interests have fractured rural communities and changed the lives of the people that live in them. Other factors have also affected the traditional life of rural communities, not least the impact of representations of the countryside as an idyllic place to live. This is the focus of Chapter 3.

CHAPTER 3

THE GOOD LIFE

The countryside is increasingly regarded as an attractive place to live. More and more people are choosing to leave urban areas and live in rural ones. The rural idyll, oulined Chapter 1, has played a role in this, as have the changing employment opportunities discussed in Chapter 2. This chapter helps you to examine the patterns, processes and causes of population change in the countryside.

Depopulation

Until 40 or so years ago, the dominant population movement in more economically developed countries was one of rural-to-urban migration. This was associated with urbanisation and people moved from rural areas to take up employment in new urban industries. Today, lack of opportunity, problems of access to housing, and poor employment prospects are still driving many, particularly young, people away from rural areas. However, these movements are often hidden because the biggest *net* (or overall) population movement is from urban to rural areas. This is known as counter-urbanisation, 'population turnaround', 'the urban-to-rural shift' or 'rural repopulation'.

The pattern of counter-urbanisation

As Figure 10 indicates counter-urbanisation has been experienced in all parts of the United Kingdom over the past 30 or 40 years. Information Box 4 charts this change in England and Wales.

Information Box 4: Population movement 1950s to 1980s

1950s
Between 1951 and 1961 Cornwall and most counties in Wales underwent depopulation as a result of a decline in agricultural and other traditional forms of employment. Conversely, the population of the Home Counties increased, particularly in Essex and Hertfordshire, indicating suburbanisation and urban spillover. Related to this, London experienced a net decline in population. During this period, the population in all other metropolitan and non-metropolitan areas grew by between 1% and 10%.

1960s
The decade between 1961 and 1971 was one of counterurbanisation. For example, Cornwall and the west of Wales experienced a net growth of population in contrast to the losses of the previous decade. Powys was the only rural county to display a net loss of population, while all other rural areas registered growth. The counties of eastern England and the Midlands significantly increased in population, while the Home Counties continued to expand rapidly. A net loss of population also occurred in other metropolitan counties and, in particular, in Merseyside and Tyne and Wear.

1970s
By 1981 there was a distinct division between the growth rates of metropolitan and non-metropolitan areas. Conurbations in the West Midlands, South Wales, Merseyside, Yorkshire, Manchester, the North East and London all experienced net population losses. The main centres of population growth switched to peripheral and intermediate areas. In particular, the Welsh Borders and eastern England underwent large increases in population while growth in the Home Counties slowed down and, in the case of Surrey, actually reversed. During the 1970s the population growth was strongest in freestanding rural areas.

1980s
This trend continued during the 1980s, but at a reduced rate. Only one county – Cambridgeshire – recorded population growth at a rate of more than 10%. Other rural areas grew between 1% and 10% during this period. In all metropolitan areas, and in Gwent, Mid and West Glamorgan and Lancashire, population declined, but at a decreasing rate.

CHAPTER 3: THE GOOD LIFE

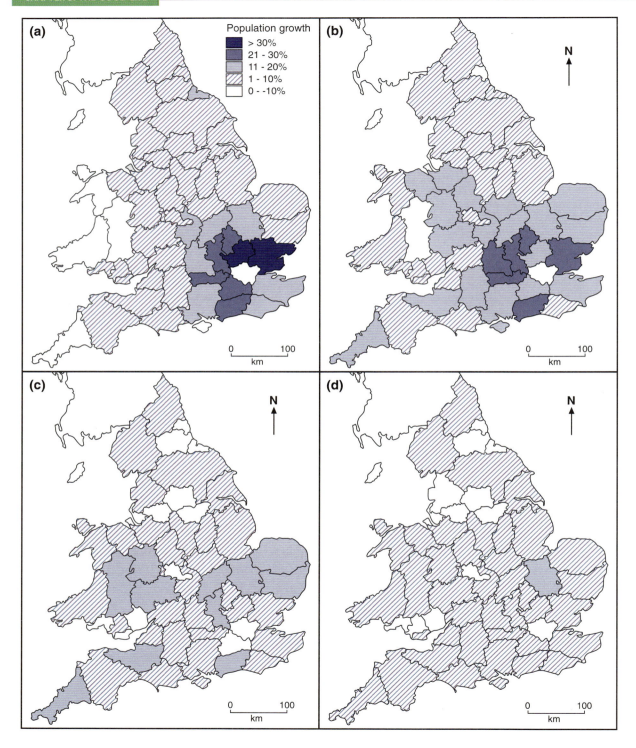

Figure 10: Net population change (percentage) by county in England and Wales: (a) 1951-61, (b) 1961-71, (c) 1971-81 and (d) 1981-91.
Source: 1991 Census, historical tables.

At the time of writing, data from the 2001 Census were not available, but they are expected to show a continuation of the counter-urbanisation trend, albeit at a slower rate.

The reasons for counter-urbanisation

There are five main reasons for counter-urbanisation. All of these have some validity, although some are more important in some areas than others (see Activity Box 8, page 28). Geography matters!

1. Changing perceptions

Chapter 1 argued that rurality is socially constructed. Over time different ideas have been held by different groups of people about the countryside. In the past, rural life may have seemed rather backward and unappealing but views have changed. The concept of the rural idyll is now the hegemonic (dominant) view of the countryside. By contrast, urban areas are seen by many as offering a poorer quality of life. As a result, many people have moved from the city to the country in search of a better lifestyle.

In a study of migration to rural Devon and Lancashire, Keith Halfacree (1994) interviewed 69 migrants from urban areas. His findings suggested that people were mainly moving for a better quality of physical environment (59% of migrants), or for a better quality of social environment (41%). Likewise, in a separate study of 338 migrants to the Scottish Highlands (Jones *et al.*, 1986), it was noted that the main reasons for moving were:

- Wanting to live in a nice area – 57%
- Employment – 24%
- To be near friends or relatives – 9%
- Other reasons - 8%
- Housing – 2%

Although movement for reasons to do with the quality of the environment may be more prevalent in remote or scenic areas, the desire to live in idyllic rural surroundings has influenced movement to all areas of rural Britain. Retirement also contributes to this movement because people who have toiled in the city all their lives equate rural living with a less stressful and more pleasant lifestyle.

The desire for a rural way of life, results in some migrants 'buying into' the rural experience. Barn conversions, for example, are popular with many urban migrants. Figure 11 shows a barn being converted into luxury homes in Shropshire. Here, purchasers are buying not just a house, but one that has a rural heritage. Such houses somehow feels more 'authentic' and lives up to the expectations of rurality of those moving to the countryside. Inside rural (and urban) homes, furnishings may reflect this rural heritage (Figure 12). In these ways ideas about the rural myth are brought to life and commodified for sale. Thus, the rural idyll is made tangible and becomes an artificial, but no less real, version of reality for many people. The more that people look for the 'genuine article', the more opportunities there are to commodify rurality and sell it to eager customers.

Figure 11: Converting a barn into luxury homes, Shropshire.

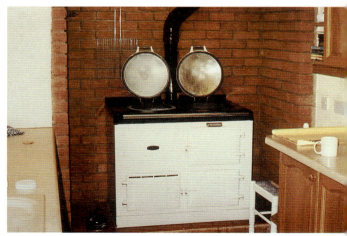

Figure 12: Owning an aga – part of the rural idyll.

CHAPTER 3: THE GOOD LIFE

2. Improved commuting and technological advances

A substantial proportion of inward migration to the countryside can be explained by commuting. As the idea of living in the countryside has become more attractive, more and more people have sought to live in the country and work in the city. In the first phase of urban-to-rural migration, commuters moved to villages in or close to the urban/rural fringe. However, improvements in the road network (including the construction of motorways, dual carriageways and by-passes) and wider access to private transport have meant that the number and range of commuters has increased considerably. Figure 13 illustrates changes to the west of the West Midlands conurbation over a 30-year period. For each district, it shows the percentage of the economically active population who work in the West Midlands (i.e. commuters).

Some geographers do not regard these movements as 'counterurbanisation' in its truest sense. This is because these commuter migrants retain close links with urban areas through their employment; thus the population movements are more akin to suburbanisation. Just as, in the past, the growth in bus and rail services pushed the suburbs further and further away from the city centre, so increased access to private transport is extending suburban life into the countryside.

The introduction of information and communications technology (ICT) has allowed some people to work from home without having to travel to work every day. This is known as teleworking or telecommuting. Instead of physically travelling to an office in the city, employees can instead work electronically: receiving, completing and sending work

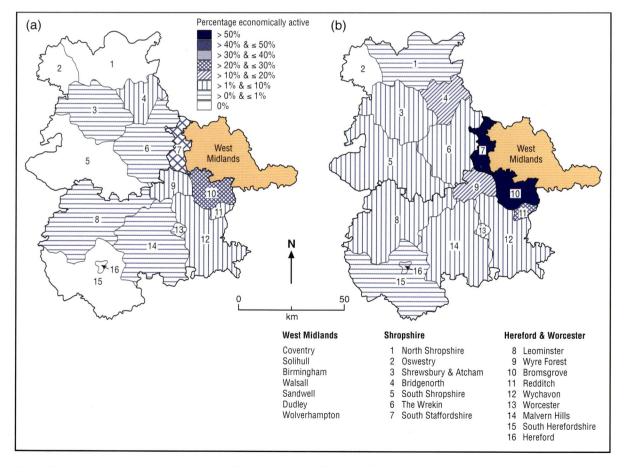

Figure 13: Commuting change (percentage) in the West Midlands: (a) 1951, and (b) 1981. Source: 1991 Census – workplace tables.

from the comfort of their own homes. Now people can live in the remotest rural region, but still work for companies based in urban areas anywhere in the world. It can be argued that, due to the developments in ICT, the distinctions between urban and rural economies are becoming even more blurred.

3. Employment change
Chapter 2 described how rural areas have undergone considerable restructuring of employment. It has been suggested that these patterns of employment have contributed to counterurbanisation. In some cases new forms of employment have led to new jobs and new migration. For example, high technology companies that locate in rural areas (e.g. 'Silicon Fen' in Cambridgeshire, or 'Silicon Glen' in Livingstone, Scotland) may bring in new workers.

Other migrants have chosen to 'downsize', perhaps giving up an old (city) job and taking a less well paid but less stressful or more rewarding one. Sometimes this might involve setting up a small business or running a smallholding, though evidence from Devon suggests that only a minority of movers have chosen this option (Bolton and Chalkley, 1989). In many cases, however, the opportunity for employment in or near a rural place can provide the impetus and finance needed to enable people to move to the countryside. Thus, a move for employment reasons may allow people to achieve their rural idyll.

4. Impact of planning and intervention
Some rural areas have been given special designation to protect the landscape and encourage recreation (see Chapter 5, pages 47-51). Consequently, these places become even more attractive to migrants who perceive such areas as less spoilt and therefore more idyllic. However, as building is extremely limited in areas with special landscape designation, any increase in in-migration increases pressure on local housing markets and reduce access to homes for local people (Chapter 4).

5. Personal circumstances
Remember that migration is complex because it is concerned with people, and people do not always behave in ways that fit the neat models devised and used by geographers. Sometimes people move for personal reasons: to be closer to family, to marry or to be closer to areas where they have 'roots'. A recent phenomenon in Britain is an increase in the number of *households.* This is for three reasons. First, many people are choosing to marry later or not at all, thus many younger people live alone. Second, people are living longer so there are more lone pensioners. Finally, the increasing divorce rate is leading to the separation of families and into two households rather than one. One consequence of these changes is that demand for housing is increasing in both urban and rural Britain.

Income also affects choice: those with high incomes have a much wider choice of where to live and it tends to be the wealthier middle-classes who are moving to rural areas. Those on low incomes can be trapped in particular places and, given high house prices in many rural areas, cannot afford to move. In addition, people on lower incomes may be forced to move out of rural areas in search of housing or employment. Thus movement is occurring up, as well as down, the urban hierarchy.

Counter-urbanisation provides an excellent topic for a questionnaire survey and, by talking to people, you can find out their reasons for moving to particular places (see Activity Box 8).

The impact of counter-urbanisation
Counterurbanisation is both a cause and effect of rural change. On the one hand it reflects changing social and economic structures and on the other, new populations can result in changes to local communities.

Social change
Migration to and from the countryside can bring important social changes to villages or small towns. In general terms, the young and the poor have left rural areas, to be replaced by wealthier, middle-class migrants. This is referred to as *gentrification*. For example, in Inkberrow, a parish in rural Worcestershire, two housing estates were built in the 1970s and, as a consequence, many people from surrounding cities moved to the village for its environment and community spirit, though they continued to commute to work elsewhere. According to the 1991 Census, the parish had a population of just under 2000 people in 700 households. Nearly 90% of homes were owner-occupied and half of the heads of households worked in professional or

managerial occupations. The majority of the parish's working population commuted to work, and 64% travelled outside the district to work. For local people, accommodation has become scarce, despite the construction of social housing in the village, and many are unable to live in or near their place of birth.

In many rural areas the age profile of the population has changed as more retired people have moved to live there and younger residents have moved out in search of work or accommodation. Some have referred to this trend, perhaps a little unfairly, as 'geriatrification'! Table 4 shows how the age profile of three Herefordshire towns changed over a ten-year period.

Activity Box 8: Investigating counter-urbanisation

Examine Figure 14, which shows adjoining rural districts and illustrates their net population change over a ten-year period.
1. Describe the main patterns of population change.
2. Using evidence from the map, explain the changes in population in each of the Districts A, B and C.
3. Discuss one social impact of the demographic changes experienced in Districts B and C.
4. Why do you think these changes have occurred?

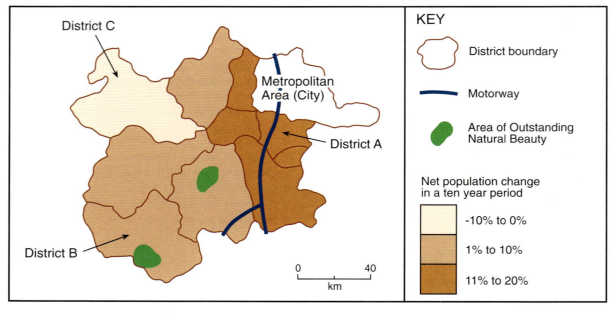

Figure 14: Population change in adjoining rural districts. After: WJEC, 2000.

Table 4: The age of residents in three Herefordshire towns, 1981 and 1991. Source: Census data.

Town/area	Percentage under 16		Percentage between 16 and pensionable age		Percentage of pensionable age	
	1981	1991	1981	1991	1981	1991
Bromyard	23	18	59	57	18	25
Kington	20	16	56	56	24	28
Leominster	22	19	58	56	20	25
Hereford and Worcester (county)	23	20	60	61	17	19

Cultural change

Culture refers to shared practices and identity. It can refer to deeply held historical identities as well as to more everyday activities and practices. It has been argued that counter-urbanisation has had an impact on local identity in the countryside. In rural Wales, for example, the use of the Welsh language declined in the twentieth century. In 1901, 50% of people living in Wales could speak Welsh but by 1991 this figure had fallen to 18.7% (though the rate of decline has now slowed and is even being reversed in some areas). Although the reasons for this decline are many and complex, an influx of English-speaking people to rural Welsh-speaking areas meant that the language was used by fewer people on a daily basis. The influx of relatively wealthy incomers also led to an increase in house prices, making it difficult for Welsh-speaking locals to find housing in their own communities at prices they could afford.

In most parts of the British countryside, much of the change that has taken place in everyday country life is attributed to 'newcomers', and local people may resent this. For example, in a survey of rural residents in Herefordshire some people who felt themselves to be 'local' made the following comments:

- 'Newcomers try to change everything.'
- 'We don't mind newcomers, so long as they don't alter things.'
- 'Newcomers have taken over.'

Looked at from a positive standpoint, new people can bring vibrancy to a village. On the one hand young families with children can help to maintain local services and, in particular, primary schools. The latter are particularly important as they provide social links for children and parents alike. The presence of a school in a village also helps to retain or attract young families to it. Retirees, especially 'young retirees', are the people most likely to volunteer for community activities or involve themselves in local societies as they have the time and experience to run these properly. Figure 15 shows the vibrancy of activities in Henley-in-Arden, Warwickshire.

In many 'gentrified' villages a key priority is conservation and preservation of the environment or heritage of the community. While this may appear altruistic, it is also about protecting property and privilege. Some commentators have suggested that there is a 'drawbridge' effect whereby people move to a village for its environment and are then determined that nobody else will follow them to spoil it! Consequently, these migrants are keen to join parish councils, village preservation societies and local conservation groups to campaign against any changes in a village that might detract from their ideal view of rurality or the values of their property.

However, it is over-simplistic to regard the split between the interests of locals and newcomers as either typical of most villages, or the main issue in terms of rural conflict. Work through the scenarios in Activity Box 9. As you will see, very few, if any, of these divisions can be explained by local or newcomer status alone. Rather, it is more helpful to consider economic and social interests.

Caution is needed when discussing cultural change. Some texts or commentators refer to 'urbanites' or 'ruralites'. These are not particularly helpful terms as they imply that a person's behaviour and identity can be explained by their upbringing in an urban or rural environment, whereas social and economic interests are more likely to determine behaviour. It is also dangerous to think of all change as being 'a bad thing'. Urban areas have experienced many social, economic and cultural changes over recent years. Although cities are not without their problems, their cultural diversity and vibrancy should be a cause for celebration. Should rural areas be any different? Increasingly, the term 'true' country person is being used to justify a rather narrow, élite vision of the countryside for the few and not the many. Less emphasis should be placed on defending a somewhat idealistic vision of rural life, and more attention given to addressing the social and economic inequalities that are typical of many rural areas.

Counter-urbanisation has brought more people to the countryside, reflecting many of the social and economic changes that have affected rural areas. However, a more rigorous analysis reveals that rural conflict and hardship cannot be attributed to, or explained away by, an influx of newcomers. Social interests, government policies and economics all play a part in creating disadvantage or opportunity in the countryside. This is examined in more detail in Chapter 4.

CHAPTER 3: THE GOOD LIFE

Activity Box 9: Constructing and contesting Bovineton: a role-play

Bovineton (fictitious) was originally an agricultural village and many farms in the parish are being forced to diversify in order to survive. The village has seen an influx of new residents, many of whom continue to commute to jobs in nearby cities. Many of the younger residents, who have grown up in the parish, are finding it difficult to find work or affordable housing.

Consider the following individuals and groups in the parish:

Farmer Brown has a mixed family farm in the parish but wants to diversify. He plans to establish a clay-pigeon shooting facility on his land that would bring in valuable income. He is also prepared to sell some land to a local housing association that will build affordable houses for local people.

Farmer Green is a relative newcomer to Bovineton. She runs an organic smallholding and wants to introduce organic pig farming on land near the village. She also wants to encourage visitors to see the pigs and 'pick their own' fruit and vegetables.

'Bovineton Concern' is an action group that has been established to conserve the appearance of the village. The group consists mainly of newcomers to the parish who want to preserve the rural charms that drew them there in the first place.

Bovineton Parish Council includes a mix of residents who have been voted onto it to represent local interests. At present, crime and safety are high on the agenda. It is thought that installing street lights in the village would address both these issues – it would discourage burglars and make the roads safer at night.

Local families are finding it hard to buy homes and many young people are moving away. They want affordable homes to be built in Bovineton for local people. The only available site is a field owned by Farmer Brown which is close to the homes of members of 'Bovetine Concern'.

'Forest Furnishings' are a national furniture chain. They intend to establish a branch (no pun intended!) in Bovineton to make furniture for their chain of city stores. Raw material and goods will be transported in and out of the parish. They see the parish as a source of cheap labour and the 'rural image' as of benefit to sales of their new 'Country Range' of pine furniture. They have purchased a redundant barn near the village and intend to apply for planning permission to convert it into a small factory unit.

Working in a group of three of four, look at the table below. For each of the above roles your group must decide if it would support, oppose or be indifferent to each of the five schemes. Once you have made a decision, insert a tick to show support, a cross for opposition and a question mark for indifference.

Individual/group	Clay-pigeon range	Organic pig-farm	Streetlights	Affordable homes	Factory unit
Farmer Brown					
Farmer Green					
Bovineton Concern					
Bovineton Parish Council					
'Local' Residents					
Forest Furnishings					

Now work with the rest of the class to establish a mock planning enquiry for each of the applications. You should first debate whether it should go ahead. When thinking about your decisions, consider the following:
- For what reasons does each individual/group support or oppose the various schemes?
- Does opposition reflect the individual/group's status as newcomer or local?
- Alternatively, do economic factors and social interests better explain the different views?
- What does this example tell you about the way in which different individuals and groups 'construct' and 'contest' the countryside?

Calendar & notices for July

Community Primary School Parents of prospective pupils of the above school are invited to join current parents at an Open Evening. The event takes place on Tuesday 3 July, commencing at 6pm. Visitors will be able to enjoy a demonstration of some of the children's musical talents, tour the classrooms and meet the teachers – parents of existing pupils will also have the opportunity to discuss their progress. All are welcome

Women's Institute. The Pastor will be telling us about his visit to Israel at our meeting on Tuesday 3 July at the Memorial Hall from 7.30pm. Visitors are always welcome. Members please remember to bring tins for our stall at the Medieval Fayre.

Parish Church Council. will meet on Wednesday 4 July in the Parish Room – 7.45 for 8.00pm. Any matters you wish to put to the PCC please inform the Rector.

Drama Society takes its first venture into Shakespeare with a presentation of 'A Midsummer Nights Dream' in the picturesque setting of the Guild Hall Garden on Wednesday 4, Thursday 5 and Saturday 7 July. Why not get together with your friends to enjoy a picnic – garden open from 6.30pm. Performance commences 7.30pm. A wine bar available and tables for 8/10 or 12 people. Tickets £8 available from The Vanity Pot.

High Bailiff's Fayre and Medieval Market will now take place on Sunday 8 July commencing at 1.00pm at the Market Cross. Given a fine day it is hoped to attract a few thousand visitors and that as many local residents as possible will join in – 'all the fun of the Fayre'. Funds raised will be used for maintenance of the Guild Hall and the Memorial Hall.

Join In and Sing! Popular songs and anthems led by the Chorus Director, City of Birmingham Symphony Orchestra on Saturday 14 July in St John's Church, Henley-in-Arden – 2.30-6.45pm. Tickets £8.00 to include the music and tea! Please contact the Rectory.

Flower Festival at Oldberrow Church on Saturday 14 – 10-5pm and Sunday 15 July 11-5pm, closing with Songs of Praise Service at 5pm. Come and see the flowers in this small country church arranged by local parishioners and join us for a Cream *tea* in a churchyard acknowledged to be one of the finest in Warwickshire for its wildflowers. Oldberrow Church is 2 miles west of Henley-in-Arden on the A4189 to Redditch at the corner of the Morton Bagot turn.

Parish Prayer Meeting on Wednesday 18 July at St Nicholas' Church at 8.00pm. Everyone welcome.

Christian Aid Week raised £1652.50 total. Many thanks for all those who gave and all who helped this collection.

Jazz in the Garden: Saturday 21 July in the Guild Hall Garden. The Ad Hoc Jazz Band will play in aid of the Memorial Hall Roof Repair Funds. Bring your own picnic – gates open from 6.30pm – Jazz from 7.30pm. Tickets £6.50, including strawberries and a glass of fizz, available from The Vanity Pot, Main Street.

Flower Club will be meeting on Friday 27 July to prepare for their 'Show day' to be held on Saturday 28 July.

Methodist Church Coffee morning will be on Saturday 14 July – 10-12 noon. Please call in if possible. Sunday Services all at 11am

Neighbourhood Watch If anyone is willing to be a co-ordinator please let the local Police Station have a note of your name and telephone number.

ADVANCE NOTICE

Barn Dance is to be held on Friday 21 September from 8.00pm until 11.00pm at Henley High School with the proceeds going to Action Replay – a local voluntary group trying to raise funds for the regeneration of the Jubilee Play Area in Henley. Tickets £8.00 adults – £5.00 children accompanied, includes a Ploughman's Supper.

Figure 15: Parish activities in Henley-in-Arden, Warwickshire. Source: Henley-in-Arden Parish newsletter, July 2001.

CHAPTER 4

BEYOND THE CHOCOLATE BOX

Rural living is not without its problems and, for many people, life is far from idyllic. In the past, social exclusion in the countryside has often been hidden because:

- the rural poor live side by side with the wealthy and, unlike the city, there are no large concentrations of deprivation;
- it is sometimes easier to 'see' deprivation in the city. Closed-down shops, graffiti, litter, boarded windows, poor urban design, dogs wandering the streets, and so on, *may* indicate an area in decline. By contrast, it is often thought that because the countryside looks pleasant, it must also be a pleasant place in which to live;
- indicators of poverty are often urban based. For example, not owning a car is sometimes taken as an indicator of poverty. However, the lack of public transport in rural areas necessitates ownership of a car. People on low incomes will run a car, but this means that they are unable to afford other household goods;
- there is sometimes an attitude that 'less of a problem means there is no problem at all'. For example, although levels of crime and drug abuse are lower in rural areas they are still a problem;
- even when poverty is acknowledged, some people hold the view that the 'close-knit' nature of rural communities means that the poor will look after each other and there is no need for structured government help.

More recently, however, rural hardship has been brought to the attention of the wider public. The foot and mouth crisis, fuel protests and countryside marches have brought rural issues into the public imagination and political spotlight. In 2001, Prince Charles stated:

> **'If the nightmare of foot and mouth disease has served any useful purpose, it is, perhaps, to have brought the desperate plight of those who live in the countryside to a far wider audience'** (HRH The Prince of Wales quoted in *The Times*, 23 July, p. 2).

After years of neglect, a series of important reports and studies into rural life have now been published. In 1990 the Church of England published *Faith in the Countryside* (Acrora, 1990), which charted the problems facing the countryside. In 1995 the Conservative Government published *Rural England: A nation committed to a living countryside* (DoE/MAFF, 1995), the first White Paper on rural affairs in England for over 30 years! This was followed in 2000 by the Labour Government's Rural White Paper *Our Countryside: The future – a fair deal for rural England* (DETR, 2000). In 1999 the Countryside Commission and the Rural Development Agency were merged to form the 'Countryside Agency', which aims to address conservation, recreation and social and economic issues in rural England. It publishes the *State of the Countryside* report every year, which is an accessible and extremely useful source of information. Even more recently the Department of Environment, Food and Rural Affairs (DEFRA) was formed from the Ministry of Agriculture, Fisheries and Food (MAFF) and sections of the Department of Environment, Transport and the Regions (DETR) to create what is, in effect, a 'ministry of rural affairs'. A Rural Task Force has also been formed by the government following the foot and mouth epidemic; this aims to re-vitalise the society and economy of rural areas affected by the disease.

As well as increased political attention, the countryside has also been brought to the public's attention through a number of vociferous rural campaign groups. The Countryside Alliance was formed in 1998, primarily to defend hunting but also to promote the interests of people living in the countryside. Its 'Countryside March' in central London on 1 March 1998 attracted over 300,000 people, making it one of the largest, and most peaceful, mass protests of recent years.

It would seem that, at last, rural problems are starting to be recognised by politicians and the public. This Chapter helps you investigate some of the challenges facing people living in the countryside.

A jigsaw of deprivation?

In the 1980s, it was realised that the countryside was facing a 'jigsaw' of deprivation. Shaw (1979) identified three inter-related problems:

1. **Household deprivation** – relating to criteria such as income and housing.
2. **Opportunity deprivation** – relating to the loss of particular facets of rural life, such as jobs and services.
3. **Mobility deprivation** – problems stemming from the inability of some rural people to gain access to jobs, services and facilities.

These concepts provide a useful introductory framework for the analysis of rural problems.

Household deprivation

Household deprivation refers to a wide range of issues related to income, housing and household goods (see Information Box 5).

Opportunity deprivation

Opportunity deprivation refers to the loss of certain facets of rural life, including services. Services have been declining steadily for years and, today, many rural communities lack what might be regarded as basic services. In 1997, the Rural Development Commission (now the Countryside Agency) survey of all the parishes in England found that:

- 41% had no permanent shop;
- 43% had no post office;
- 52% had no school;
- 29% had no village hall;
- 71% had no daily bus service;
- less than 10% had a bank, building society, a nursery, day-care centre, dentist or daily train service;
- less than 2% had a police station.

Although the rate of decline for some of these services (e.g. schools) has slowed, many rural services are in a precarious position. Even though the numbers of people living in the countryside have increased, low population densities mean that many services are unviable and uneconomic. The current rural recession also means that there is less income to spend. However, the biggest threat to local shops comes from supermarkets, especially those built in market towns, suburban areas or on the urban-rural fringe. Compared with local shops, supermarkets offer cheaper goods, more services, more ways of paying and longer opening hours, all of which are attractive to rural residents with cars and, especially, to those who work in urban areas. Small shops find it difficult to compete with supermarkets and new food regulations present problems for smaller outlets. Internet shopping may further undermine local stores although, conversely, the use of the internet may allow them to reach a wider market. Even where services continue to exist, their character may have changed. For example, rural pubs often cater for out-of-town visitors rather than local people, and many have been transformed from informal 'locals' into more formal restaurants.

Traditionally, rural areas have generally been well-served by mobile services such as mobile libraries, and delivery vans. In some villages, attempts are being made to make 'the pub the hub' of the community and to run shop and post office facilities from the pub premises (Figure 16).

Figure 16: The pub-cum-post office at St Giles on the Heath, Devon.

CHAPTER 4: BEYOND THE CHOCOLATE BOX

Information Box 5: Rural households

Income

Although, on average, households in rural wards received more income (£24,560) than those in urban wards in 2000 (£23,450) (Countryside Agency, 2001), there is a significant minority of people in the countryside who suffer from low income. In 1986, a study by Brian McLaughlin of 876 households in five areas of rural England revealed that approximately 25% of households were living on or below the poverty line (defined according to the amount of supplementary benefit claimed by households). In a repeat study published in 1994, a similar figure (23.4%) was identified (Cloke *et al.*, 1994).

One reason for this is that wages are lower in the country than the city, especially in land-based industries. In 1999, average weekly earnings were £365 in rural districts, compared with £398 in urban ones. Cornwall, at £297, had the lowest weekly earnings (Countryside Agency, 2000). Further, as Chapter 2 demonstrated, farm incomes are declining due to the current recession in farming and the reduction in demand for casual, part-time and seasonal labour, which traditionally provided much employment and income in the countryside.

Housing

Housing in rural areas is an important and emotive topic. In the early 1990s, it was estimated that 16,000 families a year were identified as homeless in rural England alone (12% of total homelessness in England). Many people have blamed 'newcomers' to rural areas for this problem, arguing that they price locals out of housing markets. This issue is part of a much wider and more complex picture. Four main reasons can be identified for the rural housing crisis: demand, supply, land and conservation.

1. Demand
- There is a high demand for housing in rural areas. This has been fuelled by migration from urban to rural areas, particularly by the middle classes who can afford to pay more for housing in the private sector. This high demand, coupled with a low rate of supply, has meant that prices have risen. According to the Countryside Agency (2001) the price of houses in rural wards were, on average, 13% higher than in urban wards.
- Rural unemployment, underemployment and low wages mean that owner-occupation is beyond the means of many local people. Clark stated that 'in many rural areas average prices are running at between four and five times the average income, which means that most local people are unable to obtain mortgages from building societies' (1990, p. 3). In many areas (such as the Lake District), the problem is further compounded by second home or holiday accommodation, that is, ownership of houses by people whose main residence is elsewhere.

2. Supply
- For many local people home ownership is out of the question due to the high costs involved. Local authorities (LAs) have found it increasingly difficult to build affordable social housing in rural areas. The 1980 Housing Act gave long-standing tenants in local-authority housing 'the right to buy' their homes. Between 1981 and 1988, 400,000 houses were transferred to owner-occupation, meaning that the supply of rented homes in rural areas fell from 39% to 26% of total stock. Between 1991 and 1997, local authorities were able only to obtain 17,000 new housing units in the countryside (Countryside Agency, 2000).
- Housing Associations (HAs) have becoming increasingly important in the provision of affordable social housing in rural areas. HAs are diverse, non-profit making organisations, which build and manage affordable housing. HAs provided 84% of new social housing in England between 1992 and 1994 (Clark, 1996). However, while the Housing Corporation identified a need for 80,000 new affordable homes between 1990 and 1995, only 17,700 were built between 1990 and 2000 (Countryside Agency, 2000). One of the main reasons for this has been the lack of cheap land (see '3. Land' below).
- The 1996 Housing Act caused concern to HAs as it extended the right-to-buy to HA tenants. However, villages with less than 3000 people were exempt from this legislation.

3. Land
- The need to preserve farmland and landscape in the countryside means that restrictions are imposed on house building. Even when land becomes available it is expensive, and the private sector is better able to afford it than social landlords (HAs and LAs).
- The 1990 Town and Country Planning Act allowed land just outside village boundaries to be released for housing as long as it is held in perpetuity by a social landlord for local people in housing need. As planning permission is not normally granted for developments in these areas, they are referred to as 'exceptions' sites and are secured by a legal agreement known as a Section 106 Agreement. The main aim of this policy is to make land available for housing and to overcome the problem of competition for land, thus reducing the cost of developing affordable housing.

4. The conservation lobby
- Concerns over affordability of and access to housing have been overshadowed, at least in popular terms, by desires to preserve the countryside from further development. This has led to a strong anti-development lobby in the countryside, illustrated by a massive rise in membership of heritage and conservationist groups.

Case study 4: Cleobury Mortimor, Shropshire

- Cleobury Mortimer is a large village/small town in southern Shropshire (its population increased by 13% between 1981 and 1991). It has a well-defined centre with over 40 retail and service outlets, including small food and comparison goods shops, pubs, a post office, bank and garages. However, despite the growth in population and housing, the number of small shops has declined in recent years. The opening of large supermarkets in nearby Ludlow (in 1988) and in Kidderminster (in 1994) had a considerable impact on trade in the town.
- Seventy per cent of households in Cleobury Mortimer and its hinterland do their main weekly shop in a supermarket. People perceive supermarkets as offering a wider range of goods, of better quality and at cheaper prices than local shops in Cleobury. Parking facilities and longer opening hours added to the attractiveness of supermarkets, especially to residents who worked outside Cleobury. A number of these commuters said that they would shop more frequently in the town if the shops stayed open after 5.30pm, that is after they returned home from work.
- An average of £24.50 per week was spent in Cleobury per household, compared to £47.00 per week in supermarkets. Over a year, an average of £2,173,000 was spent by Cleobury households in supermarkets, compared with £1,022,591 in the town itself.
- However, not all local shops were suffering. Three types could be identified. First, shops used regularly by respondents, which had a good customer base and a good 'passing' trade. These were mainly shops selling 'daily' products such as papers, cakes, bread and groceries. Second, there were those shops used little or infrequently by respondents, whose trade was declining. These were mainly those selling specialist items, such as electrical goods, shoes, clothes or hardware. Third, there was a group of outlets, particularly food shops, which appeared to have lost trade to supermarkets but which retained a loyal customer base. Significantly, there was evidence that newer specialist 'niche' businesses, such as a wine shop, were beginning to conform to this pattern and were starting to build a regular group of 'weekly' clients from within and outside Cleobury.
- Market towns, such as Cleobury Mortimer, are seeking to re-invent themselves. In 1997 'Action for Market Towns' was established to support and promote these settlements. Further, market towns have been identified by the Government as important in rural regeneration (DETR, 2000), providing a link between urban and rural areas. The Countryside Agency has a current programme to revitalise market towns in England (CA, 2001).

Source: Barrett *et al.*, 2001; CA, 2001.

Activity Box 10: Growing villages/market towns

Using information from Case Study 4 suggest:

- How individual shops could try to improve custom.
- What kind of new shops might thrive in Cleobury.
- Suggest ways that the town could be improved to attract more trade. Related to this:
  - write a marketing strategy for the town;
  - prepare a bid to LEADER (see pages 40-41) to fund improvements to the town centre.

Mobility deprivation

Mobility is a particular issue in the countryside where access to jobs, services and social activities depends on availability of transport. However, travelling adds to the cost of living in rural areas, often resulting in household deprivation. It is not surprising that the fuel protests of 2000 received support from farmers, rural pressure groups and others living in the countryside.

The rate of car ownership per household is much higher in the countryside than in urban areas – 84%

compared with 69% (DETR, 2000). For most rural households, owning a car is a necessity rather than a luxury. According to the government, 'the poorest 10% of households are twice as likely to own a car if they live in a rural area' (DETR, 2000, p. 55). But one car per household may not be enough: if the family car is used by one member of a household to travel to work others can be trapped at home. Women are more likely to be disadvantaged in this way. Dependence on private transport creates problems for other groups too: young people who cannot drive or afford private transport can be denied access to employment, services and a social life; and pensioners and others who lack access to private transport or lose the ability to drive may suffer from isolation. Public transport is the obvious alternative, but provision in many rural areas is poor (and sometimes, as Figure 17 shows, amalgamated with other services). The 2000 Rural Services Survey (CA, 2001) showed that even where bus services operated in rural England:

- only 35% of parishes had a bus operating at peak times;
- 11% had one operating between 0900 and 1500 hours;
- 59% had a bus after 1800 hours.

Many innovative measures have been taken to fill the transport gap in rural areas. For example the same survey (CA, 2001) found that:

- 39% of parishes had a Dial-a-ride scheme;
- 35% had a community taxi;
- 33% a supermarket bus.

Other recent innovations have included lending mopeds to young people for travel to work; introducing 'wiggle' buses that deviate from their standard routes to pick up passengers; and establishing 'bingo buses' to provide transport to bingo or other specific social events. In 1998 the Government's White Paper *A New Deal for Transport* outlined plans for an integrated transport policy for rural areas aimed at reducing reliance on the car, improving public transport and enhancing access to local services (DETR, 2000). Rural Bus Subsidy Grants were introduced in 1998 (worth £90 million between 1998 and 2001), to enhance existing provision and/or add new services (DETR, 2000; CA, 2001). However, many rural bus services are under-used and reliance on private cars remains high. There is a need to shift local opinion about transport use if a reliable and integrated public transport system is to be sustained in the countryside (Activity Box 11).

Figure 17: A post bus in rural Wales.

Activity Box 11: Transport provision

Obtain bus timetables and a base map for a chosen rural area. Produce public transport maps for your region. You may like to use different colours to demonstrate days of the week or time of the day that services operate.

Using your data, consider the following:

- Are any places particularly well or particularly poorly served by public transport?
- What would be the implications for people living in these areas?
- What times to buses run at? Do you consider that any of these might be problematic (e.g. no late buses back from local places of entertainment, buses running at awkward times for shopping trips)?
- Where do these buses run to? Using *Yellow Pages*, check how easy it is to access the nearest: leisure centre, hospital, doctor's surgery, cinema and other services you consider to be important in peoples' lives.

Rural 'others'

Although the concepts of opportunity, household and mobility deprivation provide a sound introductory framework to examine rural issues, some commentators have criticised the way in which social exclusion has been studied in the countryside. In particular there has been concern that rural affairs have been dominated by 'white, middle class, middle aged, able-bodied, sound minded, heterosexual men living in major urban centres of the west' (Philo 1992, p. 199). Many academics fall into these categories (the author, although not middle aged yet, is a good example!) and there has been concern that rural issues are being viewed from rather an élite perspective. Social exclusion is being defined from 'on high' and imposed on people living in the countryside.

A recent report on rural lifestyles (Cloke *et al.*, 1994) attempted to use more qualitative approaches to allow rural subjects to speak for themselves on the issues facing them. It was discovered that the same situation could be experienced differently by different people. For example, distance from services was, for some people, problematic but, for others, the isolation was a valued part of rural life. It might be argued, therefore, that it is hard to categorise somebody as suffering from 'opportunity deprivation' as the lack of shops, people and services is seen by them as a good thing.

There is a danger that some voices and experiences of rurality have been drowned out by more powerful groups. Recognising this, more effort has been made recently to listen to 'other voices' and to study minority groups in the countryside. For example, attention has been given to nomadic people (Figure 18), people of colour, children, women and people with disabilities. As a result, valuable perspectives have been gained on rural life that would otherwise have remained 'hidden'.

For example, issues surrounding race and racism have often been dismissed as unimportant in the countryside because so few people from ethnic minorities live there. In Herefordshire and Worcestershire, for example, only 1.3% of the rural population described themselves as 'non-white' in the 1991 Census, compared with 15% in neighbouring Birmingham. However, the fact that so few people of colour live in the countryside *is in itself* an issue. Julian Aygeman and Rachel Spooner (1997) suggest the following reasons for this:

- When questioned, many people from ethnic minorities said they would feel 'out of place' in the countryside. It is rare to see black faces on any images of rural life, and the countryside is generally associated with a (white) heritage in which ethnic minorities are either absent or

Figure 18: Travellers' camped in Worcestershire.

CHAPTER 4: BEYOND THE CHOCOLATE BOX

ignored. Further, because so few ethnic minority people live in the countryside or use it for recreation, others feel that they 'stand out' and may be stared at in rural areas.

- There is some evidence that more overt forms of racism can be as, if not more, problematic in the countryside than the city. The Commission for Racial Equality produced two reports highlighting this problem in 1992, one called *Keep Them in Birmingham* and the other *Not in Norfolk* – titles which apparently reflected some of the views expressed by white rural residents.
- Few people from ethnic minorities live in or use the countryside for recreation. It has been suggested that this reflects lack of income and opportunities for recreation among these people.

Because the numbers of people suffering from stress, drug abuse, homelessness, fear of crime and domestic violence are lower in rural than in urban areas, more attention has been given to these issues in the context of urban places. However, various studies of minority groups (Cloke and Little, 1998) have revealed that drug or alcohol abuse and many other 'urban' problems are as much an issue in the countryside as they are in some towns and cities.

This emphasis on diversity is known as a post-modern approach to study. Post-modernism focuses on diversity and difference, considering society as a collage rather than a grand picture. For the post-modernist the tiniest pieces of this collage are as important as the biggest ones. In contrast to a jigsaw, the pieces of a collage do not necessarily fit together perfectly! Indeed, many pieces may overlap and one person can have many identities. To research these issues, post-modernists use methods, such as interviews, that enable individuals to be studied in more depth and allowed to speak for themselves, rather than using extensive surveys that attempt to put people into particular categories. Indeed, the post-modernist geographer is likely to be more interested in the 5% of people that might answer a

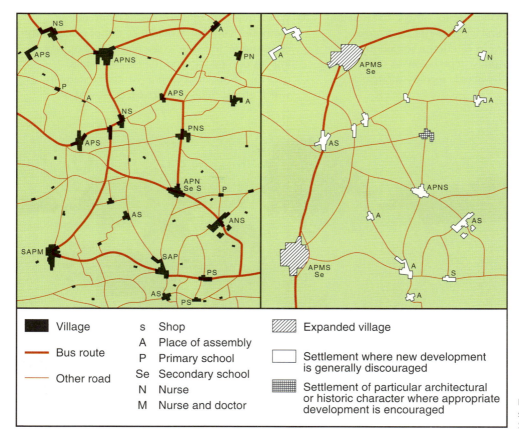

Figure 19: The key settlement principle. Source: Cloke, 1983.

survey question differently than those given by the other 95%.

What can be done?
It is easier to identify the problems facing the countryside than to solve them. However, policy makers have adopted three main strategies tackle them: top-down, bottom-up and partnership approaches.

1. The 'top-down' approach
This is where decisions are made by authorities or agencies and imposed on particular people and places. This does not mean that local people are not consulted, but it does mean that decisions are made at higher levels. The structure and local plans produced by local authorities to outline how land will be regulated through planning permission are a good example of the 'top-down' approach. Many authorities have adopted a 'key settlement' approach whereby attempts are made to cluster certain services together in places of different size and position in the graded hierarchy of settlements (Figure 19). Thus, a market town may contain a secondary school that serves a broad rural hinterland, and certain villages may be selected as sites for primary schools to serve more local populations. Grading settlements in this way ensures that most people have access to the services they need.

The advantage of the top-down approach is that it is strategic in nature. As we have seen, there are many demands and conflicts on rural space so it is helpful to have a co-ordinated plan that has been devised by professionals to take account of changes over space and time. The disadvantage of the top-down approach is that local communities may feel isolated from decision making, and while some plans may make economic sense they can be socially disastrous. In a notorious example, Durham County Council's 1951 Development Plan proposed a policy of 'planned decline' for some villages. In these cases it was suggested that 'no further investment of capital on any considerable scale should take place' (Cloke, 1983, p. 95). The idea was to channel economic development into selected villages elsewhere in the county. This fitted into a broader plan of economic regeneration for the county but took no account of the needs of those people living in the villages earmarked for 'decline'.

2. The 'bottom-up approach'
The bottom-up approach to planning is based on listening to local opinion and devising local solutions to problems rather than producing a overarching strategic plan which does not take account of local needs. The advantage is that local people are closely involved in developments and are allowed to have a say about the matters that they feel need addressing. This approach involves listening, and empowering 'other' voices, and relies heavily on voluntary effort. Some organisations act as 'enablers' to encourage local people to take part, and to co-ordinate responses.

An example of this approach was the Jigsaw/Jigso programme in rural Wales (Edwards, 1998). Launched in 1988, it aimed to 'help local communities express their needs and aspirations ... to those who might facilitate their development'. Communities were encouraged, with help from local authorities and academics, to undertake 'village appraisals' (or questionnaire surveys of all their residents) to identify local concerns. In the first five years of the campaign, 179 communities undertook an appraisal. In turn this led to many local actions, such the development of sports fields, the provision of community minibuses, or environmental schemes, to improve the quality of life in these communities.

However, 'bottom-up' approaches often lack the teeth or power to make substantial changes. They may address very local concerns but they are unlikely to influence some of the deeper issues affecting society such as lack of employment. There is also a concern that it is the more articulate, confident and experienced members of society that tend to get involved in these schemes (see Activity Box 12).

3. The partnership approach
Increasingly, a 'partnership' approach is being adopted. This aims to combine the best features of the top-down and bottom-up approaches. Partnerships are made up from representatives from the state, private and voluntary sectors; thus, a range of voices is involved. Partnerships are also well placed to draw on funding. Many use a 'matched funding' principle, whereby money may be provided from a central fund if the same amount is 'matched' by local organisations or individuals (lottery projects work on this principle). Thus, partnerships have both the resources and power to make changes.

CHAPTER 4: BEYOND THE CHOCOLATE BOX

Activity Box 12: Rural development agencies

There are several agencies with responsibility for rural development. Many have excellent websites that outline their aims, current development programmes and other innovations. Visit a selection of the following websites and find examples of development programmes.

- **European Union:** The European Union (www.europe.eu.int) is well known for its agricultural policies but it also has a number of programmes aimed at social and economic development of rural places. Between 1989 and 1999 the Objective 5b programme used structural funds to support 'the development of rural areas' in designated European regions. This has now been replaced by the Objective 1 programme that is aimed at regions, urban or rural, with a gross domestic product below 75% of the EU average. The EU has also supported the LEADER programme (www.rural-europe.aeidll.be) that aims to provide funds for small-scale rural development projects. The first two phases of this programme were aimed at designated rural areas, although the current programme (LEADER+) is less geographical in nature.
- **United Kingdom:** District and/or county councils are responsible for planning and development. Many act as 'enablers', helping other organisations and the public to develop rural areas. National Park authorities have this responsibility for national parks in England and Wales. A complete list of local authority websites and other useful organisations is available at www.tagish.co.uk. The Department of Environment, Food and Rural Affairs (www.defra.goc.uk) was created in 2001 to co-ordinate government policy on rural affairs.
- **England:** The Countryside Agency (www.countryside.gov.uk) is responsible for leisure and social affairs in the English countryside. It also funds Community Councils – county-based organisations that encourage rural communities to take action to help themselves (www.acre.org.uk). Regional Development Agencies seek to integrate strategically urban and regional development at a regional scale. Advantage West Midlands (www.advantage-westmidlands.co.uk) is one example.
- **Wales:** The Countryside Council for Wales (www.ccw.gov.uk) is equivalent to the Countryside Agency in England, while the Welsh Development Agency (www.wda.co.uk) is concerned with the economic and structural development of Wales. The National Assembly for Wales has responsibility for rural affairs in Wales (www.wales.gov.uk/subiagrculture/index.htm).
- **Scotland:** Scottish Natural Heritage (www.snh.org.uk) promotes the care and improvement of the countryside in Scotland. Highlands and Islands Enterprise (www.hie.co.uk) aims to promote structural development in the rural north (and islands) of Scotland. Devolution has also seen the establishment of the Environment and Rural Affairs Department of the Scottish Executive (www.scotlkand.gov.uk/who.depart_rural.asp).
- **Northern Ireland:** The Department of Agricultural and Rural Development for Northern Ireland (www.dani.gov.uk) is responsible for a whole range of rural affairs, including social and economic development, in Northern Ireland.

Now answer the following questions about each website you have visited:
1. Are the programmes 'top-down', 'bottom-up' or 'partnership' based?
2. Are they targeted at specific geographical areas? How are these areas decided upon?
3. What are the programmes aiming to achieve?
4. How successful have they been?
5. What do you think is good or not good about the different approaches?

The European Union (EU) LEADER programme (translated as 'links between actions for the development of the rural economy') is a good example of partnership working. LEADER was started by the EU in 1991 with the aim of regenerating local communities in the poorest rural areas of Europe. Organisations or groups in these communities can apply for funding according to a specific set of criteria (Activity Box 13). Thus, the programme has a 'top-down' strategy to ensure that funds go to the right places, for the right reasons. However, it is up to local communities and organisations to apply for these funds and to use them to address concerns that they have identified. Thus, the programme relies on a 'bottom-up' input. Local steering groups give guidance to organisations wishing to make an application and field workers are employed to facilitate the programme in local areas. LEADER is also 'matched funding': half by the EU and half by the community making the bid. LEADER has funded a number of innovative projects across Europe. Examples include farm diversification schemes, economic surveys of town centres, sustainable forestry, the promotion of local food products, establishment of craft industries and support for local carnivals or events. All of these are aimed at promoting economic and social regeneration in a sustainable and culturally sensitive manner.

Partnership working is becoming increasingly common but it is not without its problems. Because of the numbers of partners involved, some decision making may be very bureaucratic. Although the term 'partnership' implies equality, some partners may be more powerful than others and local partnerships can be dominated by the interests of more powerful groups. It remains to be seen which partnerships will succeed and which will fail to improve the quality of life in rural places.

Contrary to stereotyped assumptions about 'the good life', rural places are not problem free. Rural areas suffer from problems similar to those experienced in urban areas, but they are often not subject to the same media coverage as those in urban areas. However, such problems are starting to gain attention and a variety of efforts are now being made to resolve them.

Activity Box 13: Assessing projects for funding

Using the criteria listed above, imagine that you are a member of a local LEADER development group and decide whether or not the following projects should be funded.

Project 1: Village conference room
A parish council and the local village hall committee wants to furnish and fit a conference room in the Village Centre to enable immediate use for conferences, seminars and training courses. This will help to sustain and improve the economy of the village by providing income for the hall and to local suppliers of goods and services.

Project 2: Farm woodland development
A farmer has recently acquired a linear wood with a stream beside it and wants to develop a range of activities in the wood. The project involves the purchase of machinery to assist in maintaining the paths using woodchips produced on site, handrails for improved access and a viewing shelter at the end of the wood to provide a resting place and a hide to enable the study of birds, badgers and other wildlife.

Project 3: Local folk music
A local music society would like funding to record local folk music. A local music publisher is willing to promote and sell the recordings. The profits would be shared between performers and the music society, who would use it to encourage more young people to play folk music from the area. Funding is sought for recording equipment, studio time and the cost of publishing the recordings.

When deciding whether or not to award LEADER funding to a project, the following criteria need to be met:

- Is the applicant eligible (is it a group, partnership or association rather than an individual or single business)?
- Is the project new or innovative?
- What benefits could it bring to the community or local economy?
- Is it sustainable?
- What impacts?

Where you decide a project is not to be funded, work out how could it be re-written to win funding? (Note: All of these projects are based on real applications that have been made to LEADER.)
Source: Grenville Sheringham, Teme Valley LEADER.

CHAPTER 4: BEYOND THE CHOCOLATE BOX

Some geographers have argued that it is wrong to think that rural problems can be 'planned out' or resolved by providing services in innovative ways. They believe that these approaches focus far too much on service provision and fail to address the social and economic inequalities that underlie so many of the problems in the countryside. In other words, they tackle the symptoms rather than the causes of rural hardship. Political, social and economic inequalities are deeply engrained in the structure of society and many believe that nothing short of radical reform will improve the situation.

All too often social problems in the countryside remain hidden from the public. Instead, many people use the countryside as a 'place of play' and are unwilling or unable to appreciate the difficulties faced by rural people. Rural leisure is a two-edged sword. On the one hand rural areas are 'commodified' in order to attract to tourists, but, on the other hand, the leisure industry is an important source of income and employment. These issues are investigated in Chapter 5.

CHAPTER 5
A PLACE TO PLAY

The countryside offers wonderful opportunities for sport, leisure and recreation. Between 1993 and 2000, the number of tourist trips to the English countryside increased by 50% and, in 1999, accounted for nearly 25% of total domestic tourism. These visits make a valuable contribution to the rural economy (Chapter 2, pages 13-22) so it is important for rural places to market themselves in ways that will attract visitors. Although walking remains one of the most popular activities, accounting for nearly 20% of these visits (Countryside Agency, 2001), the range and type of activities have increased dramatically. However, rural leisure activities can conflict with the life, work and environment of the countryside and careful management is needed. This chapter examines these conflicts, and the ways that they, and the promotion and management of tourism, reveal much about rural society.

Rurality and the tourist gaze

The rise of tourism
Visiting the countryside is a very popular pastime in Britain, and according to the Countryside Agency many (45%) visits are made for no other reason than 'to enjoy the countryside'. This tradition of wishing simply to gaze on beautiful scenery, and to be in a rural environment, goes back a long way – as Case Study 5 shows. Based on observations of the Lake District made by John Urry (1995), this study also illustrates the way in which changing fashions can affect the popularity of an upland area such as this.

Tourism is a very important part of the rural economy (Chapter 2, page 13) and it is to encourage tourism that the countryside has been 'commodified' to entice visitors to 'buy into' the local rural experience. Rural life, culture and environment are often presented in a sanitised form that will appeal to tourists. Such promotion not only encourages people to visit an area, but to take it home with them on tea-towels, mugs and other souvenirs.

Heritage is also a key theme in rural place marketing. The membership of heritage or preservation groups such as the National Trust has burgeoned in recent years and, during the 1980s, there was so much interest in the past that a museum was opening at a rate of one a week (Thrift, 1989). In rural areas working farms, craft villages, steam railways, stately homes, ancient monuments and other historic buildings have become popular visitor attractions. Places with a link to the past are somehow seen as offering a more 'genuine' experience to visitors seeking to find the 'real' culture or tradition of a place. This is why much is made of 'traditional' as well as 'local' and 'natural' aspects of locally produced rural products such as Devon fudge, Kendal mint cake or Lincolnshire sausages.

Cultural heritage is also important in the marketing of rural areas, and much is made of famous writers, artists, musicians and others associated with particular places. Examples include Hardy's 'Wessex' (Dorset), Constable's Norfolk, Wordsworth's Lake District, (Case study 5 and Activity Box 14) the Brontës' Yorkshire, Elgar's Worcestershire, and so on. These authors and artists originally wrote, painted or composed music to express their own, private thoughts about a particular place. However, they are now so closely associated with these places by the public, that they are part of the reason for visiting it. Museums, homes and trails help to reinforce these links.

Place associations of this kind also apply to popular culture. For example, visitors have flocked to villages that have been used to make rural television or radio programmes. Goathland in North Yorkshire, was used for the filming of the television series *Heartbeat* and, at the peak of the series' popularity, attracted 1.1 million visitors (Fish, 1996) seeking the Aidensfield experience. One shop even changed its name to 'Aidensfield Stores' to capitalise on this. Inkberrow in Worcestershire is one of the places that the village of 'Ambridge' in *The Archers* is based on. The village pub, The Bull, has an 'Ambridge lounge' (Figure 20) with pictures of the cast on the walls, and is a popular stop for the many fans of this long-running radio series. Hadfield in Derbyshire was used to film the BBC's *The League of Gentlemen,* a dark comedy about life in a rural town.

Case study 5: The draw of the Lake District

Prior to the eighteenth century, the Lake District was viewed as a wild and inhospitable place that should be avoided. This opinion was changed by the 'Grand Tour', in which wealthy people undertook to visit those places around Europe that were famed for their beauty, healthful properties, antiquity or other particular qualities. Grand Tours inspired an interest in landscape that, in turn, provided inspiration for artists, especially those who painted in the style known as 'picturesque'. Pictures painted in this style contained three elements of a landscape – background (usually mountains), foreground (a pastoral scene), and middle-ground (often a lake, linking the other two 'grounds') – and the subject most popular with travellers to Europe was Alpine scenery. The picturesque qualities of the Lake District were 'discovered' by these artists who likened its scenery to that found in the Alps. It rapidly gained popularity as an ideal setting for picturesque landscape paintings.

Towards the end of the eighteenth, and into the early nineteenth century, the Lake District inspired many 'Romantic' poets, one of the most famous being William Wordsworth who lived there for much of his life. William, his sister Dorothy and their literary friends spent hours walking in the countryside, and it was this intimate contact with the landscape that inspired much of what they wrote. Many of Wordsworth's poems describe the landscape and people of the Lake District, and their power was such that they attracted wealthy people to experience the area's qualities for themselves. Readers of the Lakeland poets and authors who visited the area generally chose to follow suit by walking rather than riding on a horse or in a carriage. As the number of visitors and walkers increased, Wordsworth became concerned that the area would be spoilt. In his *Guide to the Lakes,* published in 1822, Wordsworth suggested that the Lake District should be made a 'national property' for the benefit of those who had 'an eye to perceive or a heart to enjoy the area'. Wordsworth's poems continue to draw visitors, especially to his home at Rydal Mount.

By the mid-nineteenth century, better transport links and improved holiday allowances for working people resulted in even greater visitor numbers to the Lakes, and marked the start of the era of mass tourism. The area continued to inspire poets, painters and writers whose works attracted yet more tourists. One of the best-known writers associated with the area is Beatrix Potter (author of the *Peter Rabbit* and *Squirrel Nutkin* stories 1900s). From the mid-twentieth century, Arthur Ransome's *Swallows and Amazons* stories (1930s) also encouraged people to visit the Lakes specifically for boating activities.

Walking has always been one of the most popular forms of recreation in the Lake District, but its popularity increased significantly following the publication of A.W. Wainwright's *Pictorial Guides to the Lakeland Fells* in the 1950s and 1960s. The *Guides* encouraged more visitors to the fells, threatening the solitude that Wainwright so valued. Some ramblers, inspired by the books, now seek to climb (or 'bag') every summit identified by the author.

Today, the Lake District has a wide range of attractions for visitors of all ages and interests. In 1951, the area was designated as a National Park in order to conserve the landscape and to manage recreation. This emphasis on protection, conservation and 'commodification' has had beneficial effects for the tourist industry, but some negative effects for local people. For example, there is a shortage of affordable housing and of employment opportunities outside the tourist industry.

Although the topography of the Lake District have not changed dramatically over the past 200 years (apart from erosion by walkers!), the way the area is regarded by society has changed a great deal. From an unknown, inhospitable wilderness, the Lakes are now 'commodified' and managed to protect the set of rural ideals that emphasise recreation and conservation. An understanding of these changing 'constructions of rurality' helps to explain the changes that have taken place over time in areas such as the Lake District, and in other upland areas in Britain.
Source: Urry, 1995.

Figure 20: The entrance to the 'Ambridge Bar and Lounge' at The Bull in Inkberrow, Worcestershire.

The town was inundated with fans, especially coach loads of young people, looking for 'local shops' and other landmarks featured in the series (www.leagueofgentlemen.co.uk/town.htm).

Farming and agriculture are widely viewed as being central to rural life or, indeed, the embodiment of rural life itself. Thus, this aspect of rural living has also been 'commodified' to appeal to tourists. Numerous farm parks and working farms have been opened to give the public a taste of farm life, and to heighten interest, farmers often stock specialist breeds of farm animals with long horns, shaggy coats or unusual colours (see Case Study 1: Hatton Country World, pages 16-17). Some food producers, such as cheese makers, also open their premises to the public to enable them to see the production process in action, and the many annual agricultural shows held in various parts of the country continue to attract widespread interest.

Heritage marketing inevitably involves a partial and sanitised version of the rural past. For example, some farm parks try to re-create the past by demonstrating traditional farming skills for visitors but they are, of course, very selective in what they choose to show. While the public may like to try their hand at milking a cow or making cheese, it is unlikely that slaughtering a pig would be quite so appealing! Similarly, the view we are given of past life in stately homes and gardens is generally very biased, focusing mostly on the grandeur of the owners' lifestyles, and revealing little of the toil and suffering of those working in the fields around them or the sculleries below. Defenders of the heritage industry argue that it is these one-sided 'commodified' versions of the past that tourists want to see. Rather than seeking reality, they are seeking what is referred to as 'hyper-reality' or artificial versions of reality. Whether tourists want a more realistic view is open to discussion. Clearly, while many

Activity Box 14: Representations of the countryside

Choose a part of the countryside that is strongly associated with a particular author, poet, artist, musician, or television or radio programme. Examine either the output of one author, poet, artist, or a range of material then answer the following questions:

- How is the area portrayed?
- What local features or qualities are celebrated?
- How does this work compare with other pieces about the area?
- To what extent does the work reflect and/or affect prevailing ideas about the countryside?
- Have ideas about the countryside changed since this work was produced?
- How or why do you think that the artist chose to portray the countryside in this manner?

You can extend this work by examining the way this area is currently marketed (Activity Box 15) and how the work of your chosen artist is used to attract visitors and promote the area.

Activity Box 15: Rural tourism: qualitative fieldwork

Many geographers have turned to qualitative sources to tell them about the way a place is socially constructed (Chapter 1) and marketed. By using 'texts', such as films, books, paintings, art, poetry, advertising, television, radio, cartoons and music, it is possible to see how certain people or organisations attempt to influence ideas and feelings about certain places. Tourism relies on places marketing themselves as unique. However, this provides a very partial view of a place and, this can hide or even contribute to social problems in the country.

Choose a rural place and examine how its image had been marketed to encourage tourism. You can use a selection of postcards, holiday brochures, information leaflets, websites, books, etc. Visitor or information centres can also be considered as 'texts' that influence public opinion and behaviour. Using your chosen texts, consider how, and how successfully, it/they portrays the area and encourages visitors. The checklist below will help you to do this (note: you will need a separate copy for each type of text you use).

What type of text are you examining?

Who is it aimed at?

What is its purpose?

What are the main ideas it is trying to convey?

Who is/is not portrayed in the texts?

Are locals included? If so who/which ones?

How are visitors portrayed?

Are these groups of people portrayed differently?

Are animals used? If so, how?

How is the landscape presented?

What natural features dominate? Why?

Does it include features made by people (e.g. stone walls, houses)?

Why are some built forms included and other excluded?

What is the relationship between people and landscape/environment?

Is the text particularly striking or unusual?

Are ideas of heritage and history important? If so, how?

How might this text have an impact on social relations or behaviour in these places?

How is 'rurality' portrayed in the text?

---

accept that the experiences they buy into are not real, they nevertheless find them entertaining and relaxing, which, after all, is the point of taking a holiday!

Management

Visitors to the countryside can cause erosion and environmental damage to the very landscapes that attract them in the first place. The needs of tourists, farming and other industries can also clash, causing conflict and making it difficult to achieve a balance between conservation, recreation and development.

One way is to designate land for particular purposes so that the various demands on it can be managed in an integrated way. Every local authority has to produce a structure plan that details how the land will be used and managed on a county-wide scale. However, it has been recognised that some areas of land need to be given special consideration because of their landscape and recreational value.

National Parks

National Parks were established under the 1949 National Park and Access to the Countryside Act with the aim of preserving and enhancing of natural beauty of designated areas and to encourage quiet recreation. There are ten National Parks in England and Wales (Figure 21) plus the Norfolk Broads has status equivalent to a national park.

Two more National Parks have been proposed to cover the New Forest (currently recognised as a heritage area) and the South Downs. There are none in Scotland, although two are being planned: one around Loch Lomond and one in the Cairngorms. National Park Authorities manage the Parks and are also responsible for social and economic development within their Parks boundaries.

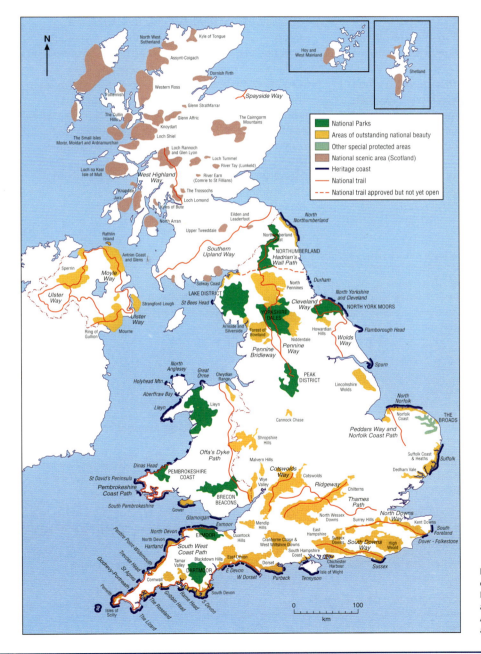

Figure 21: National Parks, Areas of Outstanding Natural Beauty, National Scenic Areas and National Trails.
After: CA, 2000; Cullingworth and Nadin, 2001.

Areas of Outstanding Natural Beauty/National Scenic Areas (Scotland)

AONBs/NSAs are in the second 'tier' of protected landscapes and were also established under the 1949 Act. However, there is no statutory responsibility to provide for recreation in these areas and they do not receive the same funding and publicity as National Parks.

Country parks

Country parks were established in 1968 to provide urban populations with a 'change of environment within easy reach'. They are the responsibility of local authorities (i.e. county and district councils) who generally provide facilities such as picnic sites, car parks, toilets, nature trails, fishing and bathing facilities, boating lakes and bridleways. Activity Boxes 16 and 17 highlight some of the issues that arise when managing a country park.

In country parks and other popular countryside sites where the habitat is less fragile, visitor centres, cafés and toilets are provided for visitors. It is often said that the vast majority of visitors to the countryside simply require a car park and a 'loo with a view'! Such visitors rarely stray more than a few metres from their car. Thus, car parks can be used to draw visitors to certain locations, thereby reducing the pressure on other areas and allowing quieter forms of recreation.

National trails

National trails, or long-distance footpaths, are examples of managed access to the countryside. Currently, there are 15 in England and Wales, three in Scotland and three in Northern Ireland (Figure 21). In addition there are many way-marked walking trails which are designated and managed by local authorities. The most popular sections of these routes may be prone to erosion but they are usually well-maintained and signposted. They bring economic benefits to adjacent areas as their users spend money in local shops, pubs and on accommodation and other attractions.

Activity Box 16: Managing a local nature reserve

Hartlebury Common (Figure 22) is owned by Worcestershire County Council and is managed by its Countryside Service as a local nature reserve. It is a 87ha (216 acre) site that attracts over 100,000 visitors a year, mainly from nearby towns and cities. Although most visitors arrive by car, many locals walk there. Of the activities that are permitted in the park, some of the most popular are:

- Dog walking
- Walking
- Horse riding
- Observing wildlife and nature
- Informal sports (football, cricket, etc.)

Clearly, some of these activities conflict with each other so the space needs to be carefully managed to ensure that all park users are able to enjoy the site.

As well as being important for recreation, Hartlebury Common is a haven for wildlife, particularly the area of lowland heath that supports scarce flora and fauna. It has been designated as a site of special scientific interest (SSSI) for these reasons. Other habitats within the park are grassland, scrub, pools and woodland, each one having a distinctive ecology.

In addition to the above the park managers have to deal with problems such as litter, dog mess, fly tipping and illegal camps, as well as with prohibited activities such as fishing, mountain biking and quad biking.

Using the above information, and bearing in mind that your budget is limited, devise a management plan for Hartlebury Common. Your plan should:

- allow a range of leisure activities to take place at the site;
- ensures that the Common's rare habitats are protected and conserved;
- include strategies for dealing with illegal or inappropriate uses of the site.

CHAPTER 5: A PLACE TO PLAY

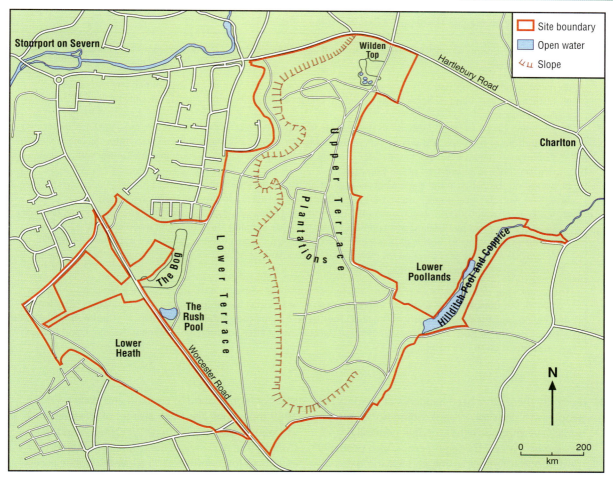

Figure 22: Hartlebury Common.

There are also a host of other designations, including sites of special scientific interest, National Nature Reserves, Marine Nature Reserves, wetland sites and Areas of Special Conservation that aim to protect fragile or rare environments, habitats and landscape from damage by recreation and development.

Despite all of these designations, these sites are difficult to manage, as there is a constant battle between the needs of conservation and recreation. Activity Box 18 helps you to appreciate some of these and to examine some of the solutions.

Access

There is an extensive network of public rights of way to the countryside of England and Wales (the total extent is 227,000km), allowing access to many areas of the countryside. The public has the right to walk or ride (on bride ways) on these routes, which should be kept accessible and open for this purpose. However, conflicts can and do arise over access along these routes. Although many landowners welcome walkers, some do not; they point out that irresponsible walkers leave gates open, trespass, destroy crops, disturb livestock and leave litter. During a recent survey, two farmers stated:

■ 'Dogs should be kept under control. People must realise that farms and the countryside are not playgrounds or picnic areas.'

49

Activity Box 17: Managing Hartlebury Common: some solutions

Worcestershire County Council has used a number of approaches to tackle the diverse issues at Hartlebury. Current measures include:

- maintaining clear footpaths that allow people to see (but not encroach on) a range of habitats and views;
- restricting horse riding to bridle paths (mainly in one area of the Common), which can only be used by people who own a license;
- providing car parks around the site to stop people from parking on the Common (note: most of these are located at the bottom of the slope and on one side of the road). The plan is that this will keep most visitors to one area of the Common as they will be reluctant to walk over the road or up the hill! In this way informal sports and recreation may be confined to one part of the Common;
- educating the public about the environment through the use of information boards;
- liaising with the local public, users of the site and interest groups via forums and opinion surveys;
- managing habitats to maintain maximum biodiversity and to conserve rare species. In some cases 'edge' habitats have been created by thinning and coppicing trees to create a 'buffer zone' between woodland and adjoining grassland or heathland. This is to provide interfaces between the two habitats, which are important for wildlife, and to provide a degree of protection for more sensitive woodland or heathland areas by using this 'buffer' zone to prevent or reduce encroachment by humans.

The Council has to strike a balance with its measure of control: it wishes to encourage people to visit and enjoy the Common, but does not wish to attract too many people, or to install buildings and facilities (e.g. toilets, litter bins) that would spoil the natural beauty of the area. They believe that if they install such facilities the Common will become a 'honey pot' for visitors, and this would increase the chances of litter pollution, environmental damage and human encroachment on to rare habitats.

- 'Recently there have been incidents of trespass by people with dogs (poaching), riding horses off bridleways and footpaths – people generally walking across and damaging crops, also probably most importantly trespassing and leaving gates open causing livestock to stray, which results in damage.'

Conversely, the Rambler's Association argues that over a quarter of public rights of way are blocked by landowners and many more are poorly signposted, discouraging access.

A particular point of contention is the 'right to roam' on areas of open land. In Scotland, where rural areas are seen to be under less pressure, this has been less of a problem than in England and Wales. There is open access to the countryside through a 'system of mutual respect' (Blunden and Curry, 1990) and, historically, there has been less need for National Parks or tightly defined rights of way. However in England and Wales there has been a longstanding conflict between landowners and those who want freedom of access to uncultivated upland areas. The establishment of National Parks in 1949 opened some areas of high ground in England and Wales. Since 1989 there have also been a series of schemes, such as Countryside Stewardship, Tyr Cymen and the Countryside Access Scheme, that have offered incentives for landowners to allow the public access to their land. A recent, and significant, piece of legislation has been The Countryside and Rights of Way Act 2000. This will give people new rights (and responsibilities) to walk over large areas of land that have been designated as 'open' by the government.

Extreme sports

Rural recreation has always contained an element of challenge. Recently new technologies have resulted in a wider range of activities, including mountain biking, white water rafting, bungee jumping, 4x4 driving and snowboarding. In these sports, participants actively engage with or even 'take on' the landscape, rather than merely looking at or travelling over it. Thus, the 'rural' is valued as a challenging wilderness, which can be 'conquered' by those with skill and courage (and the right equipment!). These also provide new challenges for countryside managers who have to ensure that the environment and other recreational

CHAPTER 5: A PLACE TO PLAY

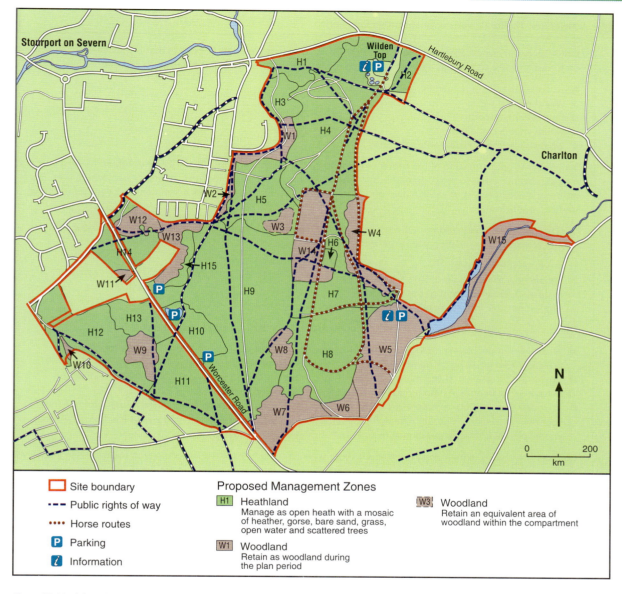

Figure 23: Hartlebury Common: access and visitor facilities and proposed management zone.

activities are not disturbed (see Activity Box 16 on Hartlebury Common).

As well as new sports, some games are extending the way in which the countryside can be used as a place of play. In Paintball games contestants are given a task, such as the capture of flags, which involves roaming over wide areas and firing paint pellets to mark their targets. If a contestant is hit in this way he/she is 'dead' and out of the game. These games are usually carried out on private (often wooded) land that has been set aside specifically for the purpose (sometimes as part of farm diversification) and are highly regulated. Despite the exhortation to 'Go wild in the countryside' (Figure 24), contestants are required to sign waiver forms beforehand, and abide by strict rules and wear safety equipment during the game. Weekend newspapers are full of adverts offering the chance to fly warplanes, drive tanks, rally

> **Activity Box 18: Rural tourism: quantitative fieldwork**
>
> Visitor surveys make excellent fieldwork activities. This activity is designed for a location that is popular with visitors. Choose such a site and design a survey (including a questionnaire), which will enable you to find out the following:
>
> - who uses the site (age, gender, ethnicity, economic status);
> - why the site is used;
> - how the site is used;
> - where visitors come from;
> - what visitors like or dislike about the site.
>
> Remember to ask permission of the owner or manager of the site you choose. They may be delighted to assist you, and may also be interested to see your results. Information about visitors is vital for the planning of rural leisure sites. Do not work alone, it is best to work in pairs and to undertake a health and safety risk assessment (ask your tutor) before starting your fieldwork.
>
> Use the information gathered from the survey, together with site and participant observations (i.e. taking part in the activity yourself and evaluating your experience in a critical way) to:
>
> 1. produce visitor profiles,
> 2. outline how the site is used,
> 3. evaluate the management of the site, and
> 4. suggest possible improvements to the management of the site.
>
> The activity also offers the opportunity to answer more detailed questions and to evaluate your own survey:
>
> - What types of people mainly visit the site? Do some social groups use it more than others? Or for different activities? Can anything be done to encourage a wider range of people to use the site?
> - How is the site marketed and managed? How is rurality presented and imagined? How might this influence behaviour? Is an organisation trying to put a particular viewpoint across?
> - How effective was your questionnaire? Is it really possible to categorise people into different user groups? Think of your own experiences of using a rural site. Is it always for the same purpose? Do you always feel the same about your experience? Does a questionnaire really get at these feelings and experiences?
>
> Consider other ways of improving or extending your work. You might have the opportunity to use extended interviews or 'focus groups' to examine these issues, particularly the more qualitative ones, in more depth. Is it possible to interview the manager of a site, or local people, or people who do not or would not use the site? Responses from such people will help you to view your findings from different angles and so will also help to put your survey in perspective. You might also use the methods in Activity Box 15 (page 46) to supplement your work.

cars or take part in survival courses in the countryside. Perhaps the rural is now being re-imagined to fulfil dreams of becoming a soldier, pilot, rally driver or adventurer!

Deviant leisure

Rural places are not always used in the way that they are intended. The countryside can offer space and privacy not always found in urban areas, especially after dark. Countryside managers also have to contend with various 'deviant activities' from prohibited sports (such as quad biking) to anti-social behaviour (such as fly tipping or littering) and 'deviant' behaviour (such as drug abuse or sexual acts in the bushes).

The management of these activities requires a careful and tactful approach. Some issues can be resolved by 'having a chat' with offenders and asking them to refrain from their activities; in other cases physical measures are taken to prevent such activities. For example, access to country parks is often restricted after dark, or facilities (e.g. toilets) are locked at night to prevent vandalism or drug taking. Obstacles (e.g. wooden posts) are often installed around car parks to prevent access to open ground by quad-bikers. Managers may also have to liaise with the police to

CHAPTER 5: A PLACE TO PLAY

Figure 24: Paintballing promotional material: 'Go wild in the countryside!' Reproduced with permission of National Paintball games.

prevent crime and disorder in rural places. Theft of or from cars in designated car parks is one of the biggest concerns and many now display notices warning visitors of these dangers. The safety of women in isolated areas is a serious, but very much neglected issue that may restrict their full enjoyment of recreational spaces.

Contested leisure

Some forms of rural leisure and recreation are highly controversial; hunting with hounds being a good example. The Countryside Alliance (2001) claim that fox hunting is practised by 170,000 people in Britain, and argue that it is an important part of countryside tradition, makes a valuable contribution to the rural economy and helps to protect farm animals and the environment. However, there is strong opposition to fox hunting, for example, a poll for *The Guardian* in 1995 suggested that 70% of people wanted it banned. Hunts have also been subject to the attention of hunt saboteurs who take direct action to disrupt or prevent them taking place. In 1997 Mike Foster, Labour MP for Worcester, introduced a private members bill to end hunting with hounds. In response, the Countryside Alliance (Figure 25) was formed and at a rally in London on 1 March 1998, an estimated 300,000 people expressed their opposition to a ban on fox hunting. Opposition from the House of Lords and lack of parliamentary time has prevented Foster's, and subsequent bills, from becoming law. However, efforts to ban the sport have revealed just how passionate its advocates and opponents can be, highlighting the fact that leisure in the countryside is a contested activity.

'Raves' (music festivals, organised illegally to take place in rural locations) also gained in popularity in the 1990s. At one of the largest, held at Castlemorton Common, Worcestershire, in 1992, up to 50,000 people participated in week-long illegal festival that caused significant damage and disruption locally. There were also concerns about people's safety at the event, and the fact that illegal drugs were being sold and used. (Go to Activity Box 19.)

Raves and hunting have some things in common: both are contentious, undertaken by a minority of people, can cause widespread disruption, reflect cultural interest and are opposed by many. The 1994 Crime and Disorder Act (Sibley, 1994) addressed both these activities by criminalising:

- raves on open land (rave music was defined as 'sounds wholly or predominantly characterised by the emission of a succession of repetitive beats'!), and
- trespass that aimed to intimidate, obstruct or disrupt anybody from taking part in any lawful activity. (This was aimed to prevent the activities of hunt saboteurs on private land).

Thus, one activity was made illegal and the status of the other was strengthened.

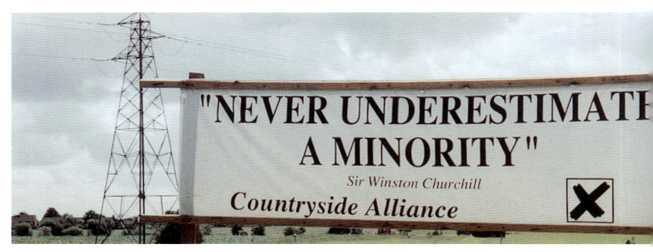

Figure 25: A Countryside Alliance banner.

Activity Box 19: Investigating rural sports

Over a period of, say, two weeks investigate a range of newspapers, magazines, television and radio programmes for advertising and articles on 'rural sports'.

- On a map of the UK, mark the location of each activity.
- Add annotations to the map, such as the type of activity and its setting (e.g. Paintballing often takes place in woodlands).
- Also add notes on how long the activity has been taking place in that location, whether it is a permanent fixture, or, where activities are temporary, their duration (e.g. Glastonbury Festival lasts a week).
- Look at the location of the activity in relation to urban areas. Do any patterns emerge? Can you say why?

As an extension you could choose one particular rural 'sport' in either one or, where possible, two or three different locations. Find out as much as you can about the history of the 'sport' in this particular location, its popularity, etc. For example, the Derbyshire Peak District has attracted climbers from across the UK because of its range of climbing edges (millstone grit and limestone) for many years, whereas paintballing is a recent sport, but a glance through the adverts in weekend papers will indicate that it is growing in popularity. You will need to gather evidence (i.e. use questionnaire surveys) both from the organisers of the sport and from the people who take part. Some of the questions from Activity Box 18 will be useful. Carefully select material from your survey and prepare a report on the activity and/or give a presentation on particular aspects of the sport.

The study of leisure in the countryside reveals how the countryside is valued and used by different groups of people, resulting in conflicts of interest and the contesting of rural space. It also reveals the many different ways that rurality is imagined. Although this is a fascinating subject for study by geographers, for those who are responsible for managing countryside leisure activities, it presents a considerable challenge and requires balance and compromise. Some would argue that the way that the countryside is managed ultimately reflects hegemonic (dominant) interests about the countryside and the desires of the most powerful groups within it. It must be remembered, however, as the examples of hunting and of the right to roam have shown, that as political, economic and social circumstances change, so does the balance of power between different interest groups.

REFERENCES AND FURTHER READING

Agyeman, J. and Spooner, R. (1997) 'Ethnicity and the rural environment' in Cloke, P. and Little, J. (eds) *Contested Countryside Cultures: Otherness, marginalisation and rurality*, London: Routledge, pp. 197-218.

Barrett, H., Storey, D. and Yarwood, R. (2001) 'From market place to marketing place: retail change in small country towns', *Geography*, 86, 2, pp. 159-62.

Blunden, J. and Curry, N. (1990) *A People's Charter? Forty years of the 1949 National Park and Access to the Countryside Act.* London: HMSO.

Bolton, N. and Chalkey, B. (1990) 'The population turnaround: a case study of North Devon', *Journal of Rural Studies*, 6,1 pp. 29-44.

Britton, D. (1990) *Agriculture in Britain: Changing pressures and policies.* Wallingford: CAB.

Champion, A., Coombes, M. and Openshaw, S. (1984) 'New regions for a new Britain', *Geographical Magazine*, 56, pp. 187-91.

Clark, D. (1990) *Affordable Rural Housing.* Cirencester: ACRE.

Clark, D. (1996) 'Identifying need and solving it/What's so special about rural areas?' in Ransley, S. (ed) *Developing Effective Responses to Rural Homelessness.* Cirencester: ACRE, pp. 3-6.

Cloke, P. (1983) *An introduction to Rural Settlement Planning.* London: Methuen.

Cloke, P. and Edwards, G. (1986) 'Rurality in England and Wales 1981: a replication of the 1971 index', *Regional Studies*, 20, pp. 289-306.

Cloke, P. and Little, J. (eds) (1998) *Contested Countryside Cultures: Otherness, marginalisation and rurality.* London: Routledge.

Cloke, P., Milbourne, P. and Thomas, C. (1994) *Lifestyles in Rural England.* London: HMSO.

Countryside Agency (2000) *The State of the Countryside 2000.* Cheltenham: Countryside Agency.

Countryside Agency (2001) *The State of the Countryside 2001.* Cheltenham: Countryside Agency.

Countryside Alliance (2001) www.countryside-alliance.org – accessed, 10 August.

Department of Environment, Food and Rural Affairs (DEFRA) (2001) 'Foot and mouth disease' defra.gov.uk/footandmouth/about/current/source.asp – accessed 7 August.

Department of Environment, Transport and the Regions (DETR) (2000) *Our Countryside: The future – a fair deal for rural England.* London: HMSO.

Department of Environment/Ministry of Agriculture, Fisheries and Food (DoE/MAFF) (1995) *Rural England: A nation committed to a living countryside.* London: DoE/MAFF.

Edwards, W. (1998) 'Charting the discourse of community action: perspectives from mid Wales', *Journal of Rural Studies*, 14, 1, pp. 63-78.

Fish, R. (1996) 'The restructuring of rural areas through media imagery' in Bowler, I. (ed) *Progress in Research on Rural Geography. Department of Geography occasional paper* 35. Leicester: Leicester University, pp. 53-4.

Gilg, A. (1996) *Countryside Planning.* London: Routledge.

Gudgin, G. (1995) 'Regional problems and policy in the UK', *Oxford Review of Economic Policy*, pp. 18-63.

Halfacree, K. (1994) 'The importance of "the rural" in the constitution of counter urbanisation: evidence from England in the 1980s', *Sociologia Ruralis*, 34, pp. 164-89.

Harrington, V. and O'Donohugue, D. (1998) 'Rurality in England and Wales in 1991: a replication and extension of the 1981 Index', *Sociologia Ruralis*, 38, 2, pp. 178-203.

Ilbery, B. and Bowler, I. (1998) 'From agricultural productivism to post-productivism' in Ilbery, B. (ed) *The Geography of Rural Change.* Harlow: Addison Wesley, pp. 57-84.

Jones, H., Caird, J., Berry, W. and Dewhurst, J. (1986) 'Peripheral counter-urbanisation: findings from an integration of census and survey data', *Regional Studies*, 20, pp. 15-26.

North, D. (1998) 'Rural industrialisation' in Ilbery, B. (ed) *The Geography of Rural Change.* Harlow: Addison Wesley, pp. 161-88.

Philo, C. (1992) 'Neglected rural geographies: a review', *Journal of Rural Studies*, 8, pp. 193-207.

Shaw, M. (1979) *Rural Deprivation and Planning.* Norwich: Geo Books.

Short, J. (1991) *Imagined Country.* London: Routledge.

Sibley, D. (1994) 'The sin of transgression', *Area*, 26, pp. 300-3.

Thrift, N. (1989) 'Images of social change' in Hamnett, C., McDowell, L. and Sarre, P. (eds) *The Changing Social Structure.* London: Sage, pp. 12-43.

Turner, M. (2001) *The FMD Epidemic in Devon: A revised assessment of its potential impacts on the county's economy.* Devon: Devon County Council.

Urry, J. (1995) *Consuming Place.* London: Routledge.

Welsh Joint Education Committee (2000) *Geography Unit 2: Urban and rural change examination*, Thursday 2 January. Cardiff: WJEC.

Yarwood, R. (1996) 'Rurality, locality and industrial change: a micro-scale investigation of manufacturing growth in the district of Leominster', *Geoforum*, 27, 1, pp. 23-37.

Yarwood, R. and Evans, N. (1999) 'The changing geography of rare livestock breeds in Britain', *Geography*, 84, 1, pp. 80-6.